D1062692

ROBOTIC OBJECT RECOGNITION USING VISION AND TOUCH

THE KLUWER INTERNATIONAL SERIES
IN ENGINEERING AND COMPUTER SCIENCE

ROBOTICS: VISION, MANIPULATION AND SENSORS

Consulting Editor

Takeo Kanade

Other books in the series:

Robotic Grasping and Fine Manipulation. M. Cutkosky.
ISBN 0–89838–200–9.

Shadows and Silhouettes in Computer Vision. S. Shafer.
ISBN 0–89838–167–3.

Perceptual Organization and Visual Recognition. D. Lowe.
ISBN 0–89838–172–X.

Three-Dimensional Machine Vision. T. Kanade.
ISBN 0–89838–188–6.

Kinematic Modeling, Identification, and Control of Robotic Manipulators.
H. Stone. ISBN 0–89838–237–8.

ROBOTIC OBJECT RECOGNITION USING VISION AND TOUCH

by

Peter K. Allen
Columbia University

KLUWER ACADEMIC PUBLISHERS
Boston/Dordrecht/Lancaster

Distributors for North America:
Kluwer Academic Publishers
101 Philip Drive
Assinippi Park
Norwell, MA 02061, USA

Distributors for the UK and Ireland:
Kluwer Academic Publishers
MTP Press Limited
Falcon House, Queen Square
Lancaster LA1 1RN, UNITED KINGDOM

Distributors for all other countries:
Kluwer Academic Publishers Group
Distribution Centre
Post Office Box 322
3300 AH Dordrecht, THE NETHERLANDS

Library of Congress Cataloging-in-Publication Data

Allen, Peter K.
Robotic object recognition using vision and touch.

(The Kluwer international series in engineering and computer science ; SECS 34. Robotics)
Bibliography: p.
Includes index.
1. Robotics. 2. Robot vision. 3. Pattern recognition systems. I. Title. II. Series: Kluwer international series in engineering and computer sciences ; SECS 34. III. Series: Kluwer international series in engineering and computer science. Robotics.
TJ211.A45 1987 629.8'92 87-17265
ISBN 0-89838-245-9

Printed in the United States of America

Table of Contents

PREFACE

This work describes a multi-sensor system for robotic object recognition tasks that integrates passive stereo vision and active, exploratory tactile sensing. The complementary nature of these sensing modalities allows the system to discover the underlying 3-D structure of the objects to be recognized. This structure is embodied in rich, hierarchical, viewpoint-independent 3-D models of the objects which include curved surfaces, concavities and holes. The vision processing provides sparse 3-D data about regions of interest that are then actively explored by the tactile sensor which is mounted on the end of a robotic manipulator. A robust hierarchical procedure has been developed to integrate the visual and tactile data into accurate 3-D surface and feature primitives. This integration of vision and touch provides geometric measures of the surfaces and features that are used in a matching phase to find model objects that are consistent with the sensory data. Methods for verification of the hypothesis are presented, including the sensing of visually occluded areas with the tactile sensor. A number of experiments have been performed using real sensors and real, noisy data to demonstrate the utility of these methods and the ability of such a system to recognize objects that would be difficult for a system using vision alone.

This work was performed in the GRASP laboratory at the University of Pennsylvania. I was fortunate enough to have watched this laboratory grow into one of the major centers for vision and robotics, and I wish to thank all its members for their help and support. In particular, I would like to thank my advisor, Dr. Ruzena Bajcsy, whose constant support and probing mind have enriched me greatly.

Financial support for this research was provided by the CBS foundation, the National Science Foundation and the Army Research Office.

Last, and most important, I wish to thank my wife Barrie for her love and support without which this work could not have been done.

ROBOTIC OBJECT RECOGNITION USING VISION AND TOUCH

CHAPTER 1

INTRODUCTION

1.1. THE PROMISE OF ROBOTICS

Robots have fascinated man for many years. The idea of an "intelligent" machine that can do tasks similar to those performed by humans has been proposed by science fiction writers and futurists and embodied in movies and toys. Over the last ten years, great strides have been made towards this goal. The decreasing cost of computing power coupled with the drive for higher productivity has led to the introduction of many robots onto factory floors. There has also been an increase in the publicity and expectations about the capabilities of these machines, which I call the promise of robotics. The promise of robotics is twofold: to create machines that can perform tasks that are currently infeasible for humans and to perform with greater accuracy, lower cost and resulting higher productivity tasks that humans now perform. The class of tasks for which robots are well suited includes those that are dangerous or in unpleasant environments (undersea, outer space), those that are boring and repetitive and that humans find unstimulating, and those requiring high precision and accuracy.

The promise of robotics has yet to be fulfilled, however. Tasks which we as humans find simple and trivial are complex and difficult for a robot to perform. A human can find an arbitrary object visually in a cluttered environment and proceed to grasp the object and move it at will, avoiding obstacles along the way and not damaging the object if it is fragile. This task is beyond the capability of most robots in use today. The majority of robot tasks currently being performed consist of pick-and-place type operations in fully known and constrained environments, where total knowledge of the relevant objects to be manipulated is assumed. These robots are pretaught a series of movements by humans that correspond to the task at hand. The movements assume no change in the real work environment from the teaching sequences. Elaborate schemes are used to recreate this static environment. In handling operations, jigs and bowl feeders are used to insure that objects to be manipulated are always presented in the same location and orientation as in the teaching sequence. Systems such as these are doomed if the object arrives in a different position or orientation, if the object is defective or if a different object appears. These robots have no way of dealing with uncertainty and in fact are subject to failure should the environment change in any way. To become more flexible and useful, robotic systems[1] need to be able to adapt to different environments and be able to reason about their environments in a precise and controlled way. Without this reasoning ability, robots simply are nothing more than fancy machine tools, hard wired for a specific application but certainly not flexible or adaptable.

Why can't robots perform as well as humans in these task domains? First, robotic sensors are nowhere near as capable as human sensors. In machine vision sensing, a small error in a digital image can have alarming consequences. Human vision, on the other hand, is extremely robust, able to tolerate noise, distortion, changes in illumination, different surface reflectance functions and changes in viewing angle. Robotic hands are crude and inflexible when compared to our multi-fingered, compliant hands. Second,

[1] The term *robotic system* is used to emphasize the system nature of robots. Typically, more than one computer and computing environment is needed for a complex robotic task.

robots have difficulty recognizing error situations, let alone coping with them. Robots are controlled by deterministic computer programs that are not able to anticipate and deal with the wide range of new and unforeseen situations that may be encountered. Third, the knowledge base of a robot is usually nothing more than a series of labeled points, precluding even rudimentary reasoning ability about the objects and tasks in its environment.

Robotic systems need to progress beyond the limited capabilities described above. The promise of robotics means that robots can work in unconstrained environments. Robots need to be able to operate outside of a specific assembly line; they need to be able to function in the home and office as well, environments that cannot be as tightly constrained as a factory. As tasks become more complex, a robotic system needs to be able to understand a dynamic, uncertain world, and to understand it through a mixture of powerful sensory processing and high level reasoning about the world.

One approach to building robots for tasks such as object recognition, grasping, manipulation and collision avoidance is to study human performance, hoping to gain insights into the interactions of human sensors, motor control and cognition. Currently, this is an active area of research, but results directly applicable to actual present day robotic tasks do not seem imminent.

A second approach, and the one advocated in this work, is to exploit existing technology. What makes this hard is that building a robotic system is interdisciplinary in nature, with ideas from computer science, mechanical engineering, electrical engineering and systems engineering needed to bring a system to fruition. Previous researchers have tended to focus on a single aspect of the problem such as building a sensor system, creating a modeling system, developing a task planner, etc. These separate subsystems have, for the most part, not been integrated into a working system. Many of the results are based on simulations and on tests run in noise-free and constrained environments. There is a large disparity between what is possible in a simulated robotics environment and the actual 3-D environment a robot will work in. This work is an attempt to bridge the gap between theoretical robotics and working systems that perform object recognition tasks in noisy, unconstrained environments. It represents the integration of many different ideas and technologies into a working system for object recognition that is synergistic in nature. It is my belief that the integration problem is the central problem in robotics; only by merging many different sensors, strategies and methods, each of which may in fact be a simple, limited system can complex behavior result.

1.2. IMPROVING ROBOTIC SYSTEM PERFORMANCE

The main purpose of this work is to improve robotic system performance for the specific task of object recognition. Object recognition is a precursor to many other important robotic tasks, including grasping, manipulation, assembly and inspection. Before we can attempt such complex robotic tasks, we need to be able to correctly recognize the relevant objects in the environment. By recognition we also mean understanding an object's position and orientation in space in a viewpoint-independent manner. The objects to be recognized are common objects found in a kitchen such as plates, bowls, mugs, pitchers and utensils. This domain was chosen because it consists of fairly complex, curved surface objects that are found in the home, an environment where the promise of robotics is great but as yet unfulfilled. In addition, this domain contains objects which are designed to be grasped and manipulated. This in fact has motivated the choice of tactile sensing as one of the multi-sensor systems that will be used to perform object recognition. Given this object domain, there are four central ideas that are incorporated into the design of this system and have been implemented in order to improve the ability of robots to work in relatively unconstrained environments. They are 1) active, exploratory sensing 2) multiple sensors 3) complex world models and 4) high level spatial reasoning, each of which is described below.

1.2.1. ACTIVE, EXPLORATORY SENSING

Many robotic tasks are attempted without sensing, assuming an absolute world model that never changes. For example, in many pick-and-place operations, the objects are always in a previously known absolute position and orientation. This approach offers little flexibility. Robotic systems need the ability to use sensory feedback to understand their environment. Work environments are not static and cannot always be adequately constrained. There is much uncertainty in the world, and we as humans are equipped with powerful sensors to deal with this uncertainty. Robots need to have this ability also. Incorporating sensory feedback into robotic systems allows non-determinism to creep into the deterministic control of a robot. There is at present much work going on in the area of sensor design for robotics. Range finders, tactile sensors, force/torque sensors, and other sensors are actively being developed. The challenge to the robotic system builder is to incorporate these sensors into a system and to make use of the data provided by them. The sensors used in this research are passive stereo vision and active, exploratory tactile sensing. An active system is one that is capable of modifying its environment and demands tight control over its actuation as opposed to passive sensors which require minimal control.

1.2.2. MULTIPLE SENSORS

Much of the sensor related work in robotics has tried to use a single sensor (typically machine vision) to determine environmental properties [1,9,16,22,27,47,48,58,64]. This can be difficult as not all sensors are able to determine many of the properties of the environment that are deemed important. For example, a vision system using 2-D projections has difficulty determining 3-D shape. A common strategy in computer vision is to try to use a single sensor to determine shape properties. Many different "shape" operators have been defined by various researchers trying to isolate separate parts of the visual system that produce depth and surface information. Examples of these are shape from texture [4,34], shape from shading [29], shape from contour [28,62,67] and shape from stereo [21,41]. A potentially promising idea is to use all of these separate shape operators together in a system that will integrate their results. Unfortunately, the operators all have different sets of constraints on the object's structure, reflectance, and illumination. The integration of these many visual operators is still not well understood. Further, the nature of most machine vision processing algorithms is that they are extremely brittle. Pushing these algorithms to extract dense depth and surface information usually introduces errors.

A much more promising approach is to supplement the visual information with other sensory inputs that directly measure the shape and surface properties of the objects in the environment. Multiple sensors can be used in a complementary fashion to extract more information from an environment than a single sensor [44,59]. If vision sensing can be supplemented with tactile information that directly measures shape and surface properties, more robust and error free descriptions of object structure can result. Chapters 3 and 4 describe the machine vision sensing used in the system and chapter 5 describes the tactile sensing algorithms.

While adding sensors to a robotic system can produce more accurate sensing, it also introduces complexity due to the added problems of control and coordination of the different sensing systems. It is difficult enough to control and coordinate the activities of a single sensor system, let alone those of a multiple sensor system. Each sensor is a distributed system with different bandwidth, resolution, accuracy and response time that must be integrated into a coherent system. Chapter 6 discusses the coordination and control of the two sensors used in this system.

Multiple sensing also raises the question of strategies for intelligent use of powerful sensors. With many ways to obtain data, some may be preferable to others and yield better results. Defining these sensing strategies is an open problem. Chapter 7 discusses strategies that are used in this research and also proposes a rule-based approach to strategy formulation that will allow the knowledge base to grow incrementally as new sensors with new capabilities are added to the system.

1.2.3. COMPLEX WORLD MODELS

If robots are to use sensory data, they have to know how this data relates to the perceived environment. Sensory data is useful only up to a point. Higher level knowledge about the world needs to be invoked to put the lower level sensory data into context. Model-based object recognition is a paradigm that allows higher level knowledge about a domain to be encoded and to assist the recognition process. Recognition has two components, a data driven or bottom-up component that supplies low level sensing primitives and a high level or top-down component that utilizes these primitives to understand a scene. At some point, low level processing is too lacking in knowledge of what is being perceived to reliably continue the recognition process. It is at this point that higher level knowledge about the domain can be effectively utilized to put the lower level information into context. In object recognition systems, this information is usually contained in models that are used to relate the observables to the actual objects. The models are abstractions of the real physical objects that try to encode important information about the object in relation to the primitives and sensing environment being used. In some sense, the model information must be computable from the sensors. It is not enough to build descriptions of objects for realistic display; the models must contain criteria that are easily accessible to facilitate efficient matching of the model to a sensed object.

Chapter 2 contains a description of a hierarchical surface and feature based model for solid objects that is well suited to the object recognition task. The model encodes rich descriptions of the geometry and topology of the objects to be recognized and is organized in a hierarchical manner to allow quick and easy access to its information. It is also structured so that matching between model and sensed observations can be done on multiple levels depending upon the requirements of the recognition process.

1.2.4. REASONING ABOUT THE WORLD

It is still not enough to have complex models and sensors that are robust. Reasoning about a complex world is necessary to be able to understand spatial relationships and geometry. This reasoning can be extremely difficult, especially if many sensors and complex models are involved. Robots cannot yet possess the deep reasoning shown by humans. However, simple reasoning about spatial and geometric relationships can help. An important component of this reasoning is to allow it to be modified easily. As new sensors and models are added, the reasoning process should be extensible to include them. This reasoning ability is perhaps the most difficult of the improvements to effect. Chapter 7 discusses the methods that are used to match the low level sensory data with the objects in the model data base and discusses approaches to verification sensing that entail high level reasoning about the object's structure encoded in the models.

1.3. SYSTEM DESCRIPTION

As mentioned above, the objects to be recognized are common kitchen items: mugs, plates, bowls, pitchers and utensils. The objects are planar as well as volumetric, contain holes and have concave and convex surfaces. These are fairly complex objects which test the modeling and recognition abilities of most existing systems. The objects are homogeneous in color, with no discernible textures. The lack of surface detail on these objects poses serious problems for many visual recognition systems, since there is a lack of potential features that can be used for matching and depth analysis. The system tries to identify each object in its environment in a viewpoint independent fashion. Chapter 8 reports results from experiments that were performed to test the system's ability to recognize 3-D object structure and correctly label each object, along with determining the orientation in space of the object.

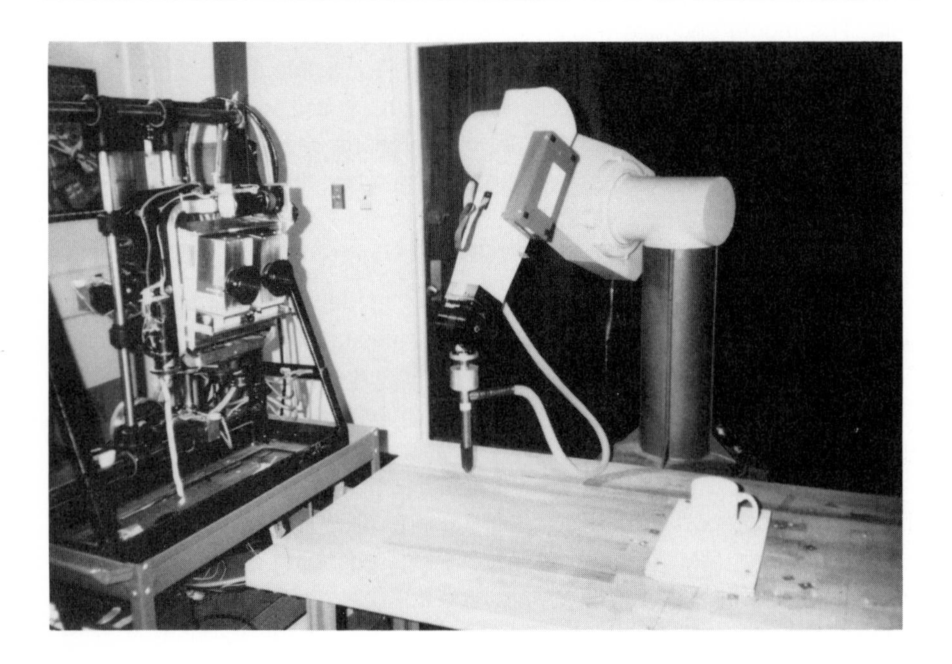

Figure 1.1. Experimental hardware.

The experimental hardware is shown in figure 1.1. The experimental procedure is to rigidly place a single object on the worktable and image it with a pair of CCD cameras. The tactile sensor is mounted on a six degree-of-freedom PUMA manipulator that receives feedback from the tactile sensor. Figure 1.2 is an overview of the software of the system. The system is divided into 6 main modules: vision sensing, tactile sensing, sensor integration, matching, verification, and the model data base.

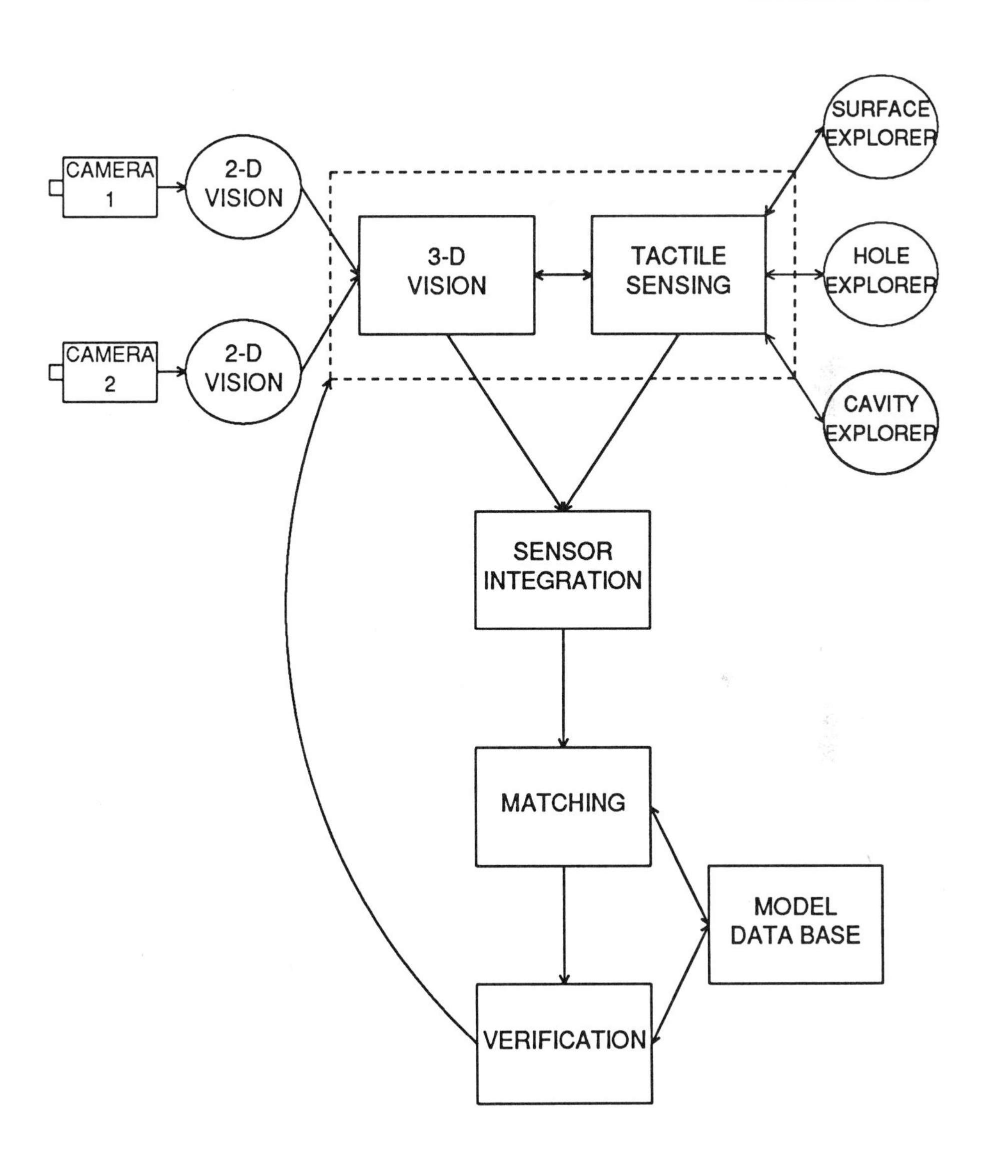

Figure 1.2. System overview.

The control flow of the system can be summarized in the recognition cycle shown in figure 1.3. This cycle is a hypothesize and test cycle, in which low level sensory information from multiple sensors (vision and touch) is integrated to form a hypothesis based upon the sensed 3-D structure of the object to be recognized. This hypothesis can then be verified by further sensing to further support or inhibit the initial hypothesis. This verification sensing is an important aspect of human recognition systems, since further sensing is essentially without cost in humans. While there is a cost in a robotic system, it is outweighed by the increased confidence in a hypothesis that verification provides. Chapter 7 discusses approaches to step 5 and chapter 8 contains an experiment that uses verification sensing to sense visually occluded areas.

---

1. The vision system images the scene and analyzes all identifiable regions of interest.
2. The tactile system explores each region identified from vision.
3. The results of the tactile and visual sensing are integrated into surface and feature descriptions.
4. The surface and feature descriptions are matched against the model data base, hypothesizing a model consistent with the sensory information.
5. The hypothesized model is verified by further active, exploratory sensing.

Figure 1.3. Recognition cycle.

1.4. SUMMARY

This work is an attempt to bridge the gap between theoretical robotics and working systems that perform object recognition tasks in noisy, unconstrained environments. The goal of this research is to make robots more flexible and adaptable, able to cope with constantly changing environments. This work extends the present capabilities of robotic systems and moves them closer to elementary reasoning about their environment. An ancillary benefit of this work is an understanding of the key problems that need to be solved to make robots more intelligent including a set of solutions for these problems in the particular task of object recognition.

Robotics is a new and changing discipline. Basic research in many areas is still under way as we try to increase our understanding of how machines may be used for complex tasks. There is a growing body of theoretical work pertaining to robotics, theory that needs to be put to use in real environments. Robotics has reached the stage where concrete examples of what robots can and cannot do are needed. There is a continuing need for a theoretical investigation of some of the difficult problems in robotic perception. However, it is also time for experimenting and implementing techniques in real, noisy and unconstrained environments.

CHAPTER 2

MODEL DATA BASE

2.1. OBJECT MODELS FOR RECOGNITION

The model data base is a key component of any recognition system. The data base encodes the high level knowledge about the objects which is needed for recognition. The global structure of the objects which is encoded in the models is used to understand and place in context the low level sensing information. The modeling system must be able to adequately model the complexity of the objects in the environment and it must contain enough structure to allow the low level sensory data to map into the model data base during the matching phase of the recognition cycle.

In deciding on a modeling system for robotic tasks, two important design decisions must be made. The first is the choice of primitive used in building the object models. Many primitives have been suggested and used for modeling 3-D objects. Badler and Bajcsy [3] and Requicha [52] provide good overviews of the different representation schemes used for 3-D objects. The choice of primitive for a model is based upon a careful analysis of the task requirements and object domain. No single representation appears to be able to adequately model all objects in all task domains. Much of the previous work in object modeling has been done in the areas of computer graphics, computer aided design (CAD) and computer vision. Many of the techniques are shared among these disciplines; however, the requirements of each modeling task tend to be quite different, affecting the choice of primitive used. Computer graphics is mainly concerned with the realistic display of objects from arbitrary viewpoints and under a variety of lighting conditions. The concern is for the final visual result rather than the underlying model's internal structure. Polygonal networks are often used in these systems for efficiency with a variety of sophisticated shading techniques utilized to create realistic displays of complex objects. The main goal of CAD systems is synthesis for the purpose of adequately creating an object for design and manufacturing purposes. Therefore, it tends to be volumetric based as an aid to the designer. Typical of this are Constructive Solid Geometry (CSG) systems such as PADL [52] and GMSOLID [10]. These systems are used to design 3-D objects by combining sets of solid primitives (cubes, cylinders, spheres, wedges, etc.) with boolean operators. Computer vision, on the other hand, tries to analyze objects for recognition. Vision systems are presented with a collection of surfaces and quadric surface and bicubic surface models are often used. A major goal in robotics is to automate the entire design and manufacturing process within one integrated system [25]. This implies the need for either an object model data base that is used for both design and recognition, or for a set of robust and efficient transformations among the different representations used. At present, no single model suffices for both tasks.

If a vision sensor is used, the second design decision is whether the object models should be 2-D or 3-D based. Image-space systems are recognition systems that try to perform recognition tasks on image properties (2-D projective properties) rather than 3-D properties. These systems are not viewpoint-independent but seek to recognize image properties derived from an number of predetermined viewpoints. Recognition occurs when one of these characteristic views is matched with an image space model of the object. Examples of this are the work of Oshima and Shirai [46,47] who used image space predictions about polyhedra and cylinders to perform recognition. Multiple learning views are computed from an object and are stored for later use. Image space curves and regions are then identified and matched with one of these views. Fisher [16] used an approach in which certain weak constraints about a surface's images over different viewpoints were computed to aid in determining the object's position and orientation. York [69] used bicubic spline surfaces as a modeling primitive and computed 2-D projective features to be used for instantiating a model.

Image space matching is not powerful because it loses the inherent sense of the 3-D object to be recognized. If the task is recognizing underlying 3-D structure, then it makes sense to model this explicitly. The projective space approach fails to maintain the consistent structure of an object across the many possible visual interpretations. The question of how many "characteristic views" of an object are sufficient is open, but clearly the answer is many. Establishing a metric on this kind of matching is difficult, especially if the sensed view is in between two stored views. 2-D projective invariants are weak, and are not robust enough to support consistent matching over all viewpoints. Koenderink [36] has developed the idea of an aspect graph that relates object geometry to viewpoint but the creation of such a graph is difficult for complex objects. What is needed is a true 3-D approach to modeling and matching, using the much stronger class of 3-D invariants.

The systems that use 3-D matching are viewpoint-independent in that matching is based upon 3-D geometric, topological and relational properties expressed in the model. [9,13,43,57]. The richer the models, the more basis for discrimination among the different objects. However, there is an extra cost associated with using a viewpoint-independent 3-D model in that they require computing a transformation from the sensed world coordinate system to the model coordinate system. This transformation can be viewed as a matrix operation with 6 degrees of freedom if the model and the imaged object are identical in size and rigid. These 6 degrees of freedom include 3 translational degrees to bring the origin of the model coordinate system into registration with the sensed coordinate system and an additional 3 degrees representing rotations around each of the 3 axes in space. These can be reduced further if the object is known to have a unique upright position. In this case 2 degrees of freedom are no longer required and a simple rotation about an upright axis is required. If scaled models are being used, then three scaling factors may also have to be computed.

2.2. CRITERIA FOR A RECOGNITION MODEL

As the previous section suggests, there is a wide range of primitives and organizations in 3-D recognition models. The design of a set of object models is intimately connected to both the task and object domain. The criteria established below are the basis for the design of the object models used in this research:

1) Computability from sensors: A model must be in some way computable from the sensory information provided by the low level sensors. If the model primitives are very different from the sensory information, then transformations which may not be information preserving are necessary. These transformations also may make the recognition process slow and inefficient. A better situation is one in which the model primitives are directly related to the low level sensing primitives.

2) Preserving structure and relations: Models of complex objects need to be broken down into manageable parts, and maintaining relationships between these parts in the model is important. In the matching process, relational information becomes a powerful constraint [13,43,58]. As the object is decomposed, it should retain its "natural" segmentation. This is important in establishing partial matches of an object.

3) Explicit specification of features: Feature based matching has been a useful paradigm in recognition tasks. If features of objects are computable, then they need to be modeled explicitly as an aid in the recognition process. Most object recognition systems are model based discrimination systems that attempt to find evidence consistent with a hypothesized model and for which there is no contradictory evidence [16]. The more features that are modeled, the better the chances of a correct interpretation.

4) Ability to model curved surfaces: Some domains may be constrained enough to allow blocks world polyhedral models or simple cylindrical objects; however, most domains need the ability to model curved surface objects. The models must be rich enough to handle doubly curved surfaces as well as cylindrical and planar surfaces. This complexity precludes many primitives, particularly polygonal networks which have simple computational properties but become difficult to work with as the number of faces increases.

5) Modeling ease: Very rich, complicated models of objects are desired. However, unless these models can be built using a simple, efficient and accurate procedure, it may be prohibitive to build large data bases of objects. Modeling is done once, so there is an acceptable amount of effort that can be expended in the modeling effort. As designs change and different versions of an object are created, incremental changes are desired, not a new modeling effort. If models are simple and easy to build, more complexity can be included in them and used for recognition.

6) Attributes easily computed: Whatever representation is used, it is important that geometric and topological measures are computed efficiently and accurately. For surfaces, this means measures such as area, surface normal and curvature. For holes and cavities this means axes, boundary curves and cross sections. Analytical surface representations such as bicubic surfaces are well suited for computing these measures.

2.3. A HIERARCHICAL MODEL DATA BASE

The criteria above has been used to build a set of models of objects for recognition tasks. Figure 2.1 shows the hierarchical model structure for a coffee mug, outlining the decomposition and structure of the models. The models are organized in a hierarchical manner which allows matching to proceed at different levels of detail, allowing for coarse or fine matching depending upon the object's complexity and the resolution of the sensing devices. It also allows for the separation of low level or bottom-up kinds of sensing from top-down or knowledge driven sensing.

The models are viewpoint-independent and contain relational information that further constrains matches between sensed and model objects. In addition, the models provide for an explicit designation of features such as holes and cavities which have proven to be powerful matching tools. Objects are modeled as collections of surfaces, features and relations, organized into four distinct levels; the object level, the component/feature level, the surface level and the patch level. The details of the model are described below.

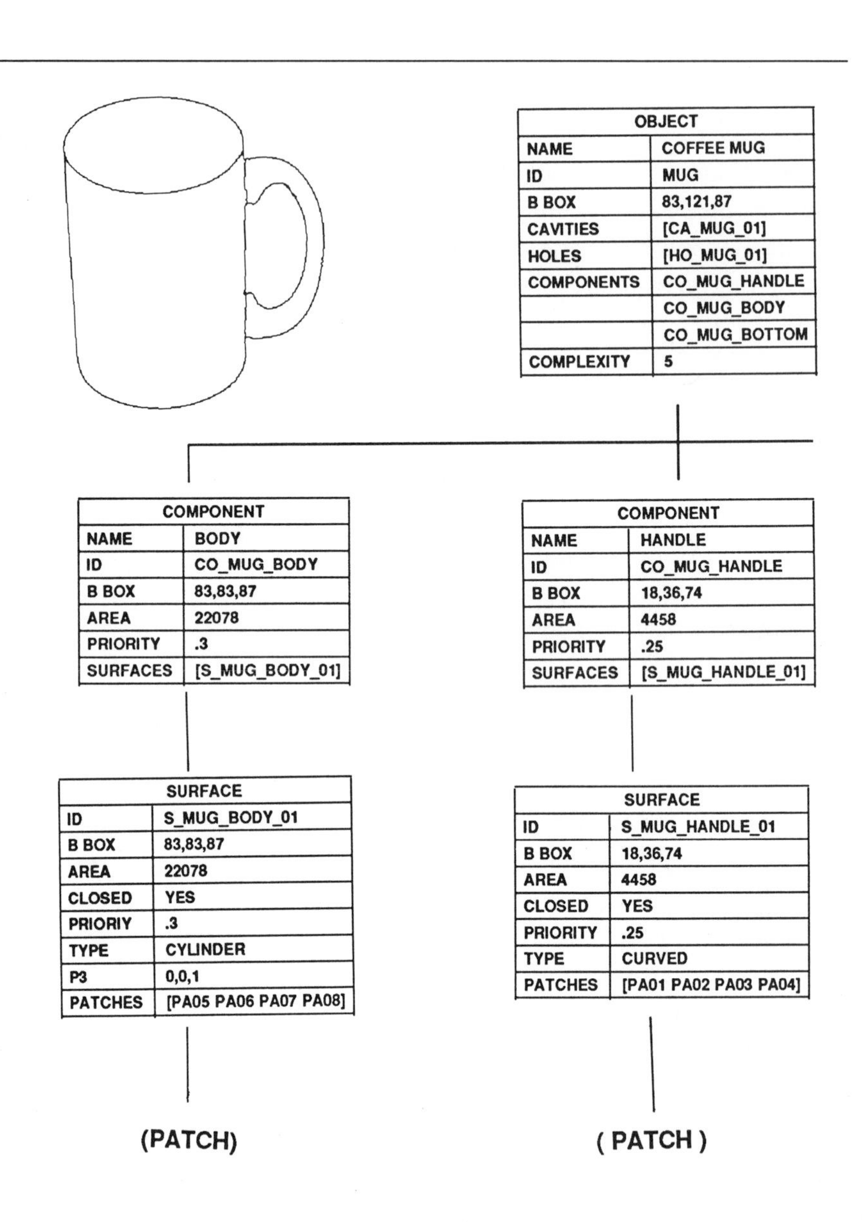

Figure 2.1a. Model structure for a coffee mug.

RELATIONS		
OBJECT	NODE1	NODE2
MUG	CO_MUG_HANDLE	CO_MUG_BODY
MUG	CO_MUG_BOTTOM	CO_MUG_BODY
MUG	CA_MUG_01	CO_MUG_BODY
MUG	HO_MUG_01	CO_MUG_BODY
MUG	HO_MUG_01	CO_MUG_HANDLE
MUG	S_MUG_BOTTOM	S_MUG_BODY
MUG	S_MUG_BODY	S_MUG_HANDLE
MUG	S_MUG_BODY	CA_MUG_01
MUG	S_MUG_HANDLE	HO_MUG_01
MUG	S_MUG_BODY	HO_MUG_01

COMPONENT	
NAME	BOTTOM
ID	CO_MUG_BOTTOM
B BOX	83,30,0
SURFACE AREA	5024
PRIORITY	.033
SURFACES	[S_MUG_BOTTOM_01]

HOLE	
ID	HO_MUG_01
AREA	1296
PRIORITY	.25
M02	187673
M20	148279
P1	0,-1,0
P2	0,0,1
P3	0,0,1
CENTROID	3,-16,249

CAVITY	
ID	CA_MUG_01
AREA	4758
PRIORITY	.166
DEPTH	87
M02	1802083
M20	1802083
P3	0,0,1
CENTROID	3,40,294

SURFACE	
ID	S_MUG_BOTTOM_01
B BOX	83,83,0
AREA	5024
CLOSED	NO
PRIORITY	.033
TYPE	PLANAR
P1	0,0,-1
P4	2.5,39.5,207
PATCHES	[PA09]

(PATCH)

Figure 2.1b. Model structure for a coffee mug.

2.3.1. OBJECT LEVEL

The top level of the hierarchy is composed of a list of all object nodes in the data base. An object node corresponds to an instance of a single rigid object. Associated with this node is a list of all the components (subparts) and features of this object which make up the next level of the hierarchy. For gross shape classification, a bounding box volumetric description of the object is included. The bounding box is a rectangular parallelepiped whose size is determined by the maximum extents of the object in the X, Y and Z directions of the model coordinate system. A complexity attribute is also included for each object. This is a measure of the number of features and components that comprise an object and it is used by the matching rules to distinguish competing matches.

2.3.2. COMPONENT/FEATURE LEVEL

The next level of the hierarchy contains two independent sets of nodes. The first set includes the components (subparts) that comprise the surfaces of the object. In the second set are the features (hole and cavities) that are used in recognition of the object. Each of these nodes is modeled differently, but they are given equal precedence in the hierarchy. They are described in detail below.

2.3.2.1. COMPONENTS

Each object consists of a number of component (subpart) nodes that are the result of a functional and geometric decomposition of an object. The components of a coffee mug are the body of the mug, the bottom of the mug and the handle. A teapot consists of a body, bottom, spout, handle and lid. These nodes are the major subdivisions of an object, able to be recognized both geometrically and functionally. Each component has an attribute list consisting of its bounding box, surface area and priority. The priority field is an aid for recognition in which the components are ordered as to their likelihood of being sensed. In the matching phase, there may be no way to distinguish between two local matches of sensed and model components. However, if priorities are included, then there is a mechanism for preferring one match over another. High priorities are assigned to large components or to isolated components that protrude in space (handles, spouts). The protruding parts may show up as outliers from the vision analysis. Obscured components, such as a coffee mug bottom when in a normal pose, are assigned lower priorities. The priority is an attempt to aid the matching probabilistically. If the object is in a regular pose, then certain parts are more prominent and can aid the matching process. Each component node contains a list of one or more surfaces that make up this functional component and that constitute the next level of the hierarchy.

The subdivision of an object by function as well as geometry is important. In some sense what determines a coffee mug is that it holds a hot liquid as well as having some familiar geometric shape. While no explicit attempt has been made here to exploit the semantic structure of objects, the model maintains a node level in the hierarchy should this be attempted. Semantic attributes as well can be hung off this node in the future to try to marry the geometric based approach with the "natural" segmentation so familiar to human beings. In most cases, the objects of the data base have a "natural" segmentation that corresponds directly with the geometry of the object. As more complex objects are modeled, this blend of functional and geometric segmentation may not be as precise.

2.3.2.2. FEATURES

Rock [53] has shown that features are important in recognition tasks for humans. If features can be recognized by sensing and matched against model features, robust recognition is possible. The features modeled in the data base are holes and cavities. Holes are modeled as right cylinders with constant arbitrary cross section occupying a negative volume. Holes can be thought of as having an approach axis which is perpendicular to the hole's planar cross section. Modeling holes as a negative volumetric entity has implications in matching. Volumetric elements have an object centered coordinate system that contains an invariant set of orthogonal axes (inertial axes). If the sensors can discover these axes, a transformation between model and world coordinates is defined which is a requirement of viewpoint-independent matching. Each hole node contains a coordinate frame that defines the hole. This frame contains a set of orthogonal axes which are the basis vectors for the frame. The hole coordinate frame is defined by the homogeneous matrix H:

$$H = \begin{bmatrix} P_{1x} & P_{2x} & P_{3x} & C_x \\ P_{1y} & P_{2y} & P_{3y} & C_y \\ P_{1z} & P_{2z} & P_{3z} & C_z \\ 0 & 0 & 0 & 1 \end{bmatrix} \quad (2.1)$$

$\mathbf{P}_1$ is the axis of maximum inertia of the hole's planar cross section.

$\mathbf{P}_2$ is the axis of minimum inertia of the hole's planar cross section.

$\mathbf{P}_3$ is the normal to the hole's planar cross section.

$\mathbf{C}$ is the centroid of the hole's planar cross section.

Besides the coordinate frame, each feature has a set of moments of order 2 that are used for matching. The computation of these moments is described in section 2.7.

Cavities are features that are similar to holes but may only be entered from one direction while holes can be entered from either end along their axis. An example is the well of the coffee mug where the liquid is poured. Cavities are modeled similarly to holes with a defining coordinate frame and moment set defined by the planar cross section of the cavity's opening. Cavities have the additional attribute of depth, which is the distance along the cavity's approach axis from the cavity's opening to the surface below.

2.3.3. SURFACE LEVEL

The surface level consists of surface nodes that embody the constituent surfaces of a component of the object. Each surface contains as attributes its bounding box, surface area, a flag indicating whether the surface is closed or not and a symbolic description of the surface as either planar, cylindrical or curved. For planar surfaces, a partial coordinate frame is described which consists of the centroid of the plane and the plane's outward facing unit normal vector. For a cylinder, the partial frame consists of the cylinder's axis. The object's surfaces are decomposed according to continuity constraints. Each surface is a smooth entity containing no surface discontinuities and contains a list of the actual surface patches that comprise it.

The particular form of bicubic surface patch that is being used in this research was originally studied by S. A. Coons and is known as a Coons' patch. The appendix contains a complete description of this primitive and it is discussed in detail in Faux and Pratt [15]. These patches have been used extensively in computer graphics and computer aided design. The patches are constructive in that they are built up from known data and are interpolants of sets of 3-D data defined on a rectangular parametric mesh. This gives them the advantage of axis independence, which is important in both modeling and synthesizing these patches from sensory data. Being interpolating patches, they are able to be built from sparse data which aids the modeling process. The most important property possessed by these patches is their ability to form composite surfaces with C^2 (curvature continuous) continuity. The object domain contains many curved surfaces which are difficult or impossible to model using polygonal networks or quadric surfaces. A bicubic patch is the lowest order patch that can contain twisted space curves on its boundaries; therefore, a complex smooth surface may be modeled as collections of bicubic patches that maintain C^2 continuity.

2.3.4. PATCH LEVEL

Each surface is a smooth entity represented by a grid of bicubic spline surfaces that retain C^2 continuity on the composite surface. Each patch contains its parametric description as well as an attribute list for the patch. Patch attributes include surface area, mean normal vector [51], symbolic form (planar, cylindrical, curved) and bounding box. Patches constitute the lowest local matching level in the system. The bicubic patches are an analytic representation that allows simple and efficient computation of surface patch attributes. They are easily transformed from one coordinate system to another by a simple matrix operation.

2.3.5. RELATIONAL CONSTRAINTS

It is not enough to model an object as a collection of geometric attributes. One of the more powerful approaches to recognition is the ability to model relationships among object components and to successfully sense them. The relational constraints among geometric entities place strong bounds on potential matches. The matching process is in many ways a search for consistency between the sensed data and the model data. Relational consistency enforces a firm criteria that allows incorrect matches to be rejected. This is especially true when the relational criteria is based on 3-D entities, which exist in the physical scene, as opposed to 2-D projective relationships which vary with viewpoint.

In keeping with the hierarchical nature of the model, relationships exist on many levels of the model. At present, there are no modeled relationships among objects since single objects only are being recognized. However, the inclusion of object relationships is an important next step in understanding more complex multiple object scenes. In particular, it should be possible to model the relationships among articulated parts, although no attempt has been made to do this.

The first level at which relational information is included is the component level. Each component contains a list of adjacent components, where adjacency is simple physical adjacency between components. The features (holes and cavities) also contain a list of the components that comprise their cross section boundary curves. Thus, a surface sensed near a hole will be related to it from low level sensing, and in a search for model consistency, this relationship should also hold in the model.

At the surface level each surface contains a list of physically adjacent surfaces that can be used to constrain surface matching. These relations are all built by hand, as the geometric modeling system being used has no way of computing or understanding this relationship. For the objects being modeled in the data base, this is at present simple to implement. However, a useful extension to this work would be to have these relations computed automatically by the modeling system itself.

The patch relations are implicit in the structure of the composite surface patch decomposition being used. Each patch is part of an ordered composite surface that contains relational adjacency automatically. Thus, each patch's neighbors are directly available from an inspection of the composite surface's defining knot grid.

2.4. CREATING THE MODELS

The models have been created by a combination of hand and computer modeling techniques. Initially, each object was digitized by a Polhemus 3-D digitizer. Each surface of the object was sampled as coarsely as necessary to allow the spline surfaces to be built accurately. The spline surfaces themselves were built by using the sparse surface data as input to a CAD/CAM surface modeler that produced a modified form of Coons' patch. The coefficients produced by this system were then scaled to reflect the true geometry of the surface being modeled. The output of the surface modeling for a particular surface is a knot set that defines a series of rectangular grids. Each of the grids contains coefficients for a single patch, and C^2 continuity is maintained across the patches that comprise a single surface. (Coons' patches are described in detail in the appendix) Figure 2.2 shows the surfaces that were generated from modeling a plate, a pitcher and a coffee mug. The plate consists of one surface containing 25 patches. The pitcher is made from 24 patches on the handle and 18 on the body. The mug has 4 patches on the body and 24 on the handle. In addition, the mug and pitcher each have the hole in their respective handles and cavities in their main body surface modeled as features.

2.5. COMPUTING SURFACE ATTRIBUTES

Once the surface patches are built, attributes of the patches must be calculated. A feature of the bicubic patches is that they are a true analytic representation of a surface, which allows simple calculation of the necessary attributes. The patches are parameterized in two dimensions, u and v, and can be represented in matrix form as

$$\mathbf{P}(u,v) = U\ A\ V \tag{2.2}$$

$$\mathbf{P}(u,v) = \begin{bmatrix} 1 & u & u^2 & u^3 \end{bmatrix} \ [\,A\,] \ \begin{bmatrix} 1 & v & v^2 & v^3 \end{bmatrix}^T \tag{2.3}$$

where A is a matrix of coefficients described in the appendix (equation A.19). The area of the surface can be calculated as:

$$Area = \int_0^1\!\!\int_0^1 |\,G\,|^{1/2}\, du\ dv \tag{2.4}$$

where G is the first fundamental form matrix defined as:

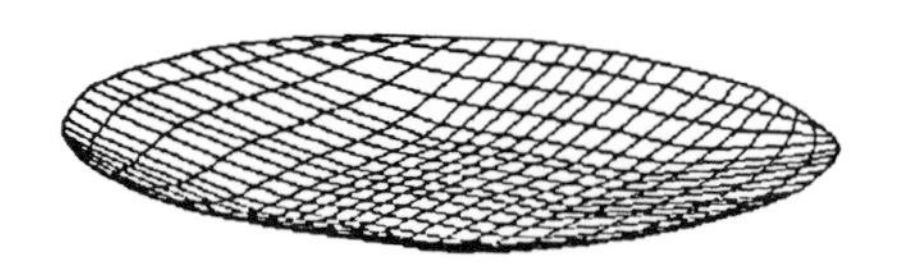

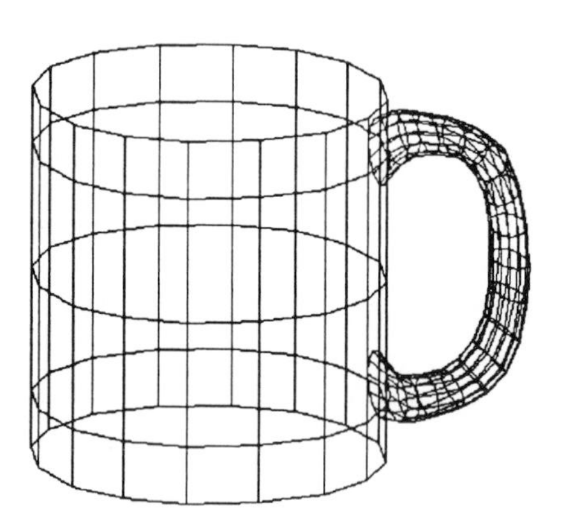

Figure 2.2. Modeled surfaces of a plate, coffee mug and pitcher.

$$G = \begin{bmatrix} \frac{\partial \mathbf{P}}{\partial u} \cdot \frac{\partial \mathbf{P}}{\partial u} & \frac{\partial \mathbf{P}}{\partial u} \cdot \frac{\partial \mathbf{P}}{\partial v} \\ \frac{\partial \mathbf{P}}{\partial v} \cdot \frac{\partial \mathbf{P}}{\partial u} & \frac{\partial \mathbf{P}}{\partial v} \cdot \frac{\partial \mathbf{P}}{\partial v} \end{bmatrix} \tag{2.5}$$

The unit normal **n** at a point on the surface can be calculated as the cross product of the tangent vectors in each of the parametric directions:

$$\mathbf{n} = \frac{\frac{\partial \mathbf{P}}{\partial u} \times \frac{\partial \mathbf{P}}{\partial v}}{\left| \frac{\partial \mathbf{P}}{\partial u} \times \frac{\partial \mathbf{P}}{\partial v} \right|} \tag{2.6}$$

The bounding box of a patch can be found analytically by finding the maxima and minima of the patch extents and subdividing the patch until it becomes planar [37]. However, this requires solving a series of equations that are cubic in one parameter and quadratic in the other, requiring numerical solution. As an alternative to the numerical solution, the surfaces were sampled at small intervals in parameter space and maximum and minimum extents were computed.

2.6. CLASSIFYING SURFACES

Differential geometry is a study of surface shape in the small, focusing on local properties of surfaces. It also provides a method of classification of surfaces by their curvature properties that can be used for matching. The Coons' patch formulation is excellent for this approach since it is an analytical form that can readily compute these curvature measures; computing such measures from point sets or polygonal approximations is difficult and error prone.

The measure that needs to be computed is the surface curvature on the patch. For a curve, curvature is well defined as

$$\kappa = 1/r \tag{2.7}$$

where r is the radius of curvature. For a surface, matters are less clear. At a single point, the curvature changes as a function of the direction moved on

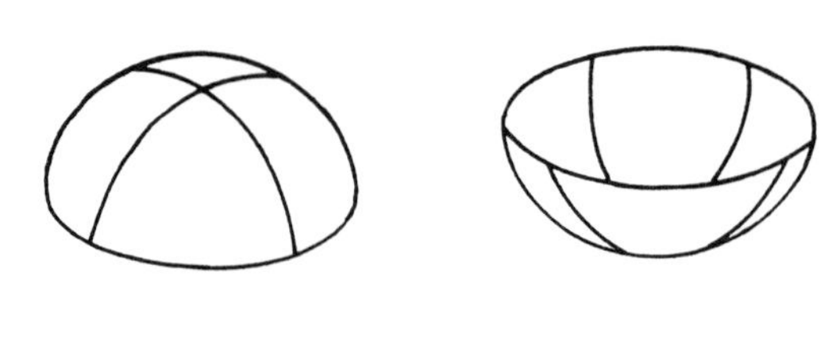

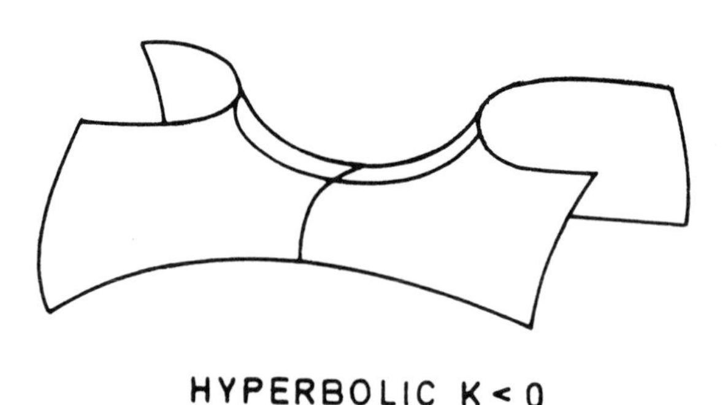

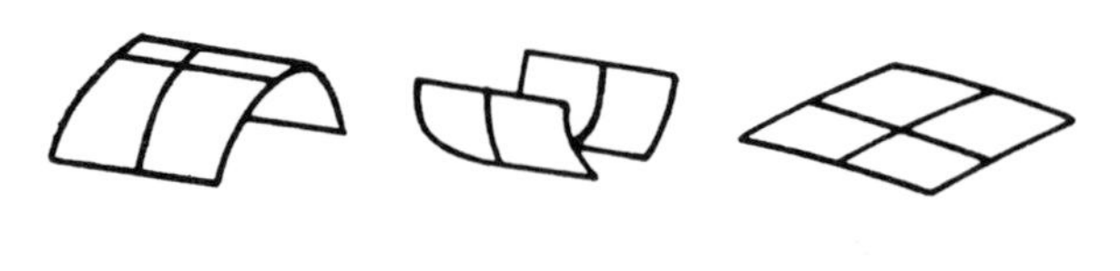

Figure 2.3. Surfaces classified by Gaussian curvature.

the surface. Limiting our discussion to regular surfaces where there is a well defined tangent plane at every point on the surface, the *normal sections* on a surface are the curves formed by the intersection of that surface with planes containing the surface normal. The curvature measured on these curves is the *normal curvature* or κ_n. As the planes containing the normal are rotated around it, forming different normal sections, different values of κ_n are defined. The directions on the surface (measured in the tangent plane) at which κ_n takes on its minimum and maximum values are referred to as the *principal directions* on the surface and define the maximum and minimum normal curvature, κ_{max} and κ_{min}. The Gaussian curvature K is defined as

$$K = \kappa_{max} \cdot \kappa_{min} \tag{2.8}$$

The Gaussian curvature is a measure that describes the local surface changes by means of a scalar (figure 2.3). Of particular importance is the sign of K. If K=0, then the curvature in one of the principal directions is zero, implying a flat surface with no curvature in this direction. It can be shown that any surface with zero Gaussian curvature can be formed by a smooth bending of the plane [26]. Planes have $\kappa_{max} = \kappa_{min} = 0$ everywhere on their surface. Cylinders also have K=0, as one of their principal curvatures is zero. A point on a surface with $K > 0$ is referred to as an elliptic point. At this point the surface lies entirely on one side of the tangent plane since both normal curvatures are of the same sign. A hyperbolic point has $K < 0$ and the surface at this point both rises above and falls below the tangent plane. By analyzing the surface's Gaussian curvature everywhere, a surface can be classified as planar, cylindrical or curved. The procedure to do this iterates over the parametric surface at a specified sampling increment, computing κ_{max}, κ_{min} and K at each point. The normal curvatures κ_{max} and κ_{min} are computed by solving the quadratic equation

$$|G|\,\kappa_n^2 - (g_{11}d_{22} + d_{11}g_{22} - 2g_{12}d_{12})\,\kappa_n + |D| = 0 \tag{2.9}$$

where G is the first fundamental form matrix defined in equation (2.5) and D is the second fundamental form matrix, defined as:

$$D = \begin{bmatrix} \mathbf{n}\cdot\dfrac{\partial^2 \mathbf{P}}{\partial u^2} & \mathbf{n}\cdot\dfrac{\partial^2 \mathbf{P}}{\partial u \partial v} \\ \mathbf{n}\cdot\dfrac{\partial^2 \mathbf{P}}{\partial v \partial u} & \mathbf{n}\cdot\dfrac{\partial^2 \mathbf{P}}{\partial v^2} \end{bmatrix} \tag{2.10}$$

2.7. COMPUTING HOLE AND CAVITY ATTRIBUTES

Features such as holes and cavities are created from the output of the surface modeler. A hole or cavity is surrounded by a series of surfaces and these boundary curves are obtained from the patch descriptions. Once the boundary points on the cross section of a hole are computed, a series of programs are run to compute inertial axes of the planar cross sections. The inertial axes are computed by finding the eigenvectors of the following matrix [54]:

$$\begin{bmatrix} M_{20} & M_{11} \\ M_{11} & M_{02} \end{bmatrix} \tag{2.11}$$

where M_{02}, M_{11}, M_{20} are the central moments of the enclosed planar cross section. The moments for a planar area are defined as:

$$M_{ij} = \int\int_{region} x^i y^j \; dx \; dy \tag{2.12}$$

Central moments are moments taken around the centroid of the object, where the centroid of a planar region is defined as:

$$\bar{x} = \frac{M_{10}}{M_{00}} \tag{2.13}$$

$$\bar{y} = \frac{M_{01}}{M_{00}} \tag{2.14}$$

These moments are computed by transforming the 3-D planar points into a 2-D plane and then using line integrals around the boundary of the cross section to compute area and moments. To transform a set of 3-D planar points into the XY plane, the coordinate frame T that describes the planar set of points in 3-D must be defined as:

$$T = \begin{bmatrix} N_x & O_x & A_x & P_x \\ N_y & O_y & A_y & P_y \\ N_z & O_z & A_z & P_z \\ 0 & 0 & 0 & 1 \end{bmatrix} \tag{2.15}$$

where **N, O, A** represent the frame's basis vectors in the reference frame and **P** represents the location of the new origin in the reference frame. The planar points normal vector will be **A**, and **N** is found by taking the vector between any two points on the plane. **O** is simply $\mathbf{A} \times \mathbf{N}$. **P** is chosen as any point in the planar point set. To transform this frame into the XY plane the frame's inverse is calculated, where the inverse is defined as:

$$T^{-1} = \begin{bmatrix} N_x & N_y & N_z & -\mathbf{P}\cdot\mathbf{N} \\ O_x & O_y & O_z & -\mathbf{P}\cdot\mathbf{O} \\ A_x & A_y & A_z & -\mathbf{P}\cdot\mathbf{A} \\ 0 & 0 & 0 & 1 \end{bmatrix} \tag{2.16}$$

Applying this transformation to the 3-D points will bring them into the XY plane. This will yield a set of planar points from which the central moments and principal axes can be computed.

Line integrals around the contour of a hole or cavity can be used to calculate the moment set. This contour is formed by linking the boundary points in a series of line segments. Beginning with Green's theorem in the plane [8]:

$$\oint_{contour} \left[P\,dx + Q\,dy \right] = \int\int_{region} \frac{\partial Q}{\partial x}\,dx - \frac{\partial P}{\partial y}\,dy \tag{2.17}$$

the centroid of the area can be computed as:

$$\bar{x} = \frac{\oint_{contour} -x\, y\, dx}{\oint_{contour} x\, dy} = \frac{M_{10}}{M_{00}} \tag{2.18}$$

$$\bar{y} = \frac{\oint_{contour} x\, y\, dy}{\oint_{contour} x\, dy} = \frac{M_{01}}{M_{00}} \tag{2.19}$$

Similarly, the moments of the enclosed area are found by:

$$M_{11} = \oint_{contour} \frac{-x\, y^2}{2}\, dx = \int\int_{region} xy\, dx\, dy \tag{2.20}$$

$$M_{20} = \oint_{contour} \frac{x^3}{3}\, dy = \int\int_{region} x^2\, dx\, dy \tag{2.21}$$

$$M_{02} = \oint_{contour} \frac{-y^3}{3}\, dx = \int\int_{region} y^2\, dx\, dy \tag{2.22}$$

Once the eigenvectors of the matrix in (2.11) are calculated, the principal axes of the cross section are found. These axes are in the plane and they must be transformed back to three space by frame T. The cross product of these transformed principal axes vectors is the hole's approach axis vector and is the normal to the set of planar contour points. Once the principal axes have been transformed to 3-D, a frame can be created with the principal axes, their cross product and the centroid as the embedded coordinate frame of the hole or cavity. This frame can then be stored in the data base and used for calculating the transformation from the sensed coordinates to model coordinates.

2.8. EXAMPLE MODEL

The models have been implemented as a series of Prolog [14] facts. The choice of Prolog for the data base was motivated by two concerns. The first was the desire to have rich relational information about adjacent parts of the model and the ability to index into the data base in many different ways. The low level sensing provides many pathways and avenues into the data base, and it is advantageous to have the model indexed by multiple levels, features and attributes. A key insight into the recognition process is that it cannot be ordered ahead of time [2]. The sensors are capable of providing different surface or feature information depending upon viewpoint. Therefore, all recognition avenues should be open at all times. Secondly, the strategies for recognizing objects are subject to change and modification. Implementing these strategies as rules is important so that the recognition behavior can be followed and modified easily. Prolog's major drawback is its lack of efficiency. For the size of the data base used in this research this posed no serious problems. As the number of objects increases, however, more powerful and faster indexing methods will be needed. Figure 2.4 is a set of Prolog facts that constitute the data base for a coffee mug. The facts include the attributes of each level as well as the relational information between entities. Rules for matching against these facts are discussed in chapter 7.

```
Object data structure:
  obj(id,bound_box,list of cavs.,list of holes,list of components,complexity)
Component data structure:
  comp(id, bound_box, surface area, priority, list of surfaces).
Surface data structure:
  surf(id,bound_box,surface area, priority,closed surface,
              kind of surface,transform).
Cavity data structure:
  cav(id,area,moment set,depth,priority,transform
Hole data structure:
  hole(id,area,moment set,priority,transform)
Relations for adjacency:
  rel(object,element1,element2)
/*******************************************************************/
obj(mug,bbox(83,121,87),[ca_mug_01],[ho_mug_01],
      [co_mug_handle,co_mug_body,co_mug_bottom],complex(5)).
cav(ca_mug_01,area(4758),mom(1802083,1802083),depth(87),pri(0.166),
      [vec(4,3,40,294),vec(3,3,40,293)]).
hole(ho_mug_01,area(1296),mom(187673,148729),pri(0.25),
    [vec(4,3,-16,249),vec(1,3,-16,249),vec(2,3,-17,249),vec(3,4,-16,249)]).
comp(co_mug_handle,bbox(18,36,74),area(4458),pri(0.25),[s_mug_handle_01]).
comp(co_mug_body,bbox(83,83,87),area(22078),pri(0.3),[s_mug_body_01]).
comp(co_mug_bottom,bbox(83,83,0),area(5024),pri(0.033),[s_mug_bottom_01]).

surf(s_mug_handle_01,bbox(18,36,74),area(4458),pri(0.25),closed,curved,[]).
surf(s_mug_body_01,bbox(83,83,87),area(22078),pri(0.3),closed,cylinder,
      [vec(4,2.5,39.5,250.5),vec(3,2.5,39.5,251.5)]).
surf(s_mug_bottom_01,bbox(83,83,0),area(5024),pri(0.033),open,planar,
      [vec(4,2.5,39.5,207),vec(3,2.5,39.5,206)]).
rel(mug,co_mug_handle,co_mug_body).
rel(mug,co_mug_body,co_mug_bottom).
rel(mug,ca_mug_01,co_mug_body).
rel(mug,ho_mug_01,co_mug_handle).
rel(mug,ho_mug_01,co_mug_body).
rel(mug,s_mug_handle_01,s_mug_body_01).
rel(mug,s_mug_body_01,s_mug_bottom_01).
rel(mug,ca_mug_01,s_mug_body_01).
rel(mug,ho_mug_01,s_mug_handle_01).
rel(mug,ho_mug_01,s_mug_body_01).
```

Figure 2.4. Prolog facts for model of a coffee mug.

CHAPTER 3

2-D VISION

3.1. INTRODUCTION

Machine vision research has been spurred by the ease with which biological systems process visual inputs. Unfortunately, the task of understanding a scene from machine vision alone has proved to be difficult. The analogy of an image matrix to a human retina has only served to illuminate the powerful kinds of processing taking place in the visual cortex, processing that is poorly understood at present. The research of David Marr and others has tried to isolate those parts of human visual information processing that seem to operate independently, such as stereopsis, and to apply this knowledge to machine vision systems. While some progress has been made, the state of machine vision is still primitive. At present, most commercial machine vision systems are binary systems that use simple template matching of 2-D silhouettes. If the object is presented in a different pose or the lighting is such that a specularity or reflection upsets the silhouette algorithms, recognition becomes impossible. What these systems lack is a way of inferring and understanding the 3-D structure of the objects to be recognized. The human visual system has little trouble performing such tasks. We can understand and recognize the objects in a scene in the presence of noise and distortion and under a variety of different lighting conditions. We can even perceive 3-D from photographs and paintings which are inherently 2-D. The as yet unaccomplished goal of machine vision systems is to to perceive as humans can.

The vision processing described here is an attempt to take what is useful and reliable from machine vision and to supplement it with active, exploratory, tactile sensing. There is no attempt to try to understand the full structure of an object from vision alone, but to use low and medium level vision processing to guide further tactile exploration, thereby invoking a consistent hypothesis about the object to be recognized. The vision processing consists of two distinct phases. The first phase is a series of 2-D vision routines that are performed on each of the stereo images. The second phase is a stereo matching process that yields sparse depth measurements about the object. The output of these modules is combined with active, exploratory, tactile sensing to produce hypotheses about objects. This chapter describes the 2-D vision processing routines in detail and discusses their performance on the images of the objects to be recognized. The next chapter discusses the stereo matching based on the output of the 2-D image processing algorithms.

3.2. IMAGE ACQUISITION

The images in this research are acquired from two 380 x 488 pixel Fairchild CCD cameras (figure 3.1) which are mounted on a movable camera frame with 4 degrees of freedom (x, y, pan, tilt). The images used here are all generated from a static camera position; no attempt was made to acquire images from multiple viewpoints. The object to be recognized is known to be a single object and in the field of view of each camera. To simplify determining figure from ground, the objects are placed on a homogeneous, black background. The lighting consists of the overhead fluorescent room lights and a quartz photographic lamp to provide enough illumination for the CCD elements.

Figure 3.1. Stereo cameras.

3.3. THRESHOLDING

The first algorithm that is run on the images is a histogram of gray levels that is used to separate out the background. Since the background is known to be homogeneous, a peak in the histogram is found that corresponds to the background gray level which predominates in the image. The picture is then thresholded at this level, driving all background pixels to zero. This gain in contrast between background and figure is helpful in establishing gradients for the object's contour.

3.4. EDGE DETECTION

Once the picture has been thresholded, an edge detection procedure is applied to both images. The edge detector that is used is the Marr-Hildreth operator, described in [40]. This operator is a derivative based operator, seeking to find intensity changes in the image array. It is defined as the convolution of the original image with the Laplacian of a Gaussian, defined as:

$$\nabla^2 G(x,y) = \frac{1}{\sigma^2}\left[\frac{x^2+y^2}{\sigma^2} - 2\right] e^{-\frac{x^2+y^2}{2\sigma^2}} \tag{3.1}$$

where σ is the standard deviation of the Gaussian and is the space constant used to determine over what scale the image should be blurred. The constant σ can be related to the image space by the formula

$$\sigma = \frac{w}{2\sqrt{2}} \tag{3.2}$$

where w is expressed as the width in pixels of the filter's central region. The idea of the Gaussian blur function is to smooth the image without destroying the underlying intensity changes. The blur function destroys all changes at a scale smaller than σ. The Laplacian is used because it is an isotropic operator, allowing a single convolution to be used that will yield orientation information. Determining the width of the filter becomes important in detecting changes at different scales. A small value of w will isolate many edge elements, while a large value of w acts as a low pass filter, allowing only large scale changes to be output.

The idea of tracking image changes from fine to coarse detail is appealing; however, it is burdensome computationally [68]. Convolving each image with the Laplacian of the Gaussian is an expensive operation, especially when performed at different scales. In this research, a single filter width was used to decrease processing time. In most cases, a small filter was used to isolate as many changes as possible, rather than miss some by using larger values of w.

The Edge Detector (algorithm 3.1) outputs the location, magnitude and orientation of each detected edge element that corresponds to a zero-crossing of the filtered image's second derivative. In any discrete approximation to this zero-crossing, the question of localization becomes important, especially if the zero-crossing locations are to be used for stereo matching. If a sign change in the convolved image occurs between two pixels in the x or y directions, a linear interpolation is used to isolate the zero-crossing to subpixels. The algorithm will find zero-crossings of both edges and noise elements in the image. A magnitude threshold is established to filter out noise edges that are of small magnitude, leaving the edge elements related to physical effects in the image. It is important to note that these physical effects include shadow, occlusions and textures as well as surface geometry.

Figure 3.2 shows the zero-crossings after applying the Edge Detector algorithm to the image of a pitcher. The zero-crossings in this image have been thresholded at the peak in the histogram of zero-crossing magnitudes, which typically lies just above zero. The zero-crossings below threshold are weak noise points and are effectively filtered by the threshold.

3.5. SEGMENTATION

Segmentation is used to isolate and analyze groups of pixels that are bounded by closed chains of edge pixels. The segmentation is used to guide the tactile system. We do not want to blindly grope on the object with the tactile sensor; we want to explore regions of interest that can be related to physical edge effects on the object. Segmentation accomplishes this goal, segmenting the object into closed contour regions that can be explored independently by the tactile system. The importance of these regions is that they are bounded by edge elements and in turn do not contain any edge elements in the interior of the region. This forms a segmentation of the object that can be used to discover the object's structure. The regions isolated on the object are either surfaces, holes or cavities which the vision system cannot determine from the sparse data available. However, the tactile exploration will be able to determine this when it begins the tactile portion of the sensing.

---

1. Convolve image P with the $\nabla^2 G$ operator, yielding image P'.
2. Proceeding left to right, top to bottom in image P', determine if a zero-crossing exists at pixel [x,y] by the following rules: Given a pixel of value A at location [x,y] in image P', surrounded by 4-neighbors of value B, C, D, E:

	$y-1$	y	$y+1$
$x-1$		E	
x	D	A	C
$x+1$		B	

a) if $(A \cdot B) < 0$ and $(A \cdot C) < 0$ then a zero-crossing exists at $[x + \mathit{interpolate}(A, B)\ ,\ y + \mathit{interpolate}(A, C)]$.

b) If $(A \cdot B) < 0$ and $(A \cdot C) \geq 0$ then a zero-crossing exists at $[x + \mathit{interpolate}(A, B)\ ,\ y]$.

c) If $(A \cdot B) \geq 0$ and $(A \cdot C) < 0$ then a zero-crossing exists at $[x, y + \mathit{interpolate}(A, C)]$.

The function $\mathit{interpolate}(a, b)$ where a and b differ in sign returns a value between 0 and 1 based on the linear interpolation of the zero point between a and b.

3. The magnitude and orientation of the zero-crossing at [x,y] is:

$$dx = P'(x+1, y) - P'(x-1, y)$$

$$dy = P'(x, y+1) - P'(x, y-1)$$

$$\mathit{Magnitude} = \sqrt{dx^2 + dy^2}$$

$$\mathit{Orientation} = \mathit{atan2}(dy, dx)$$

4. If the magnitude is below threshold M, reject this edge.

Algorithm 3.1. Edge Detector.

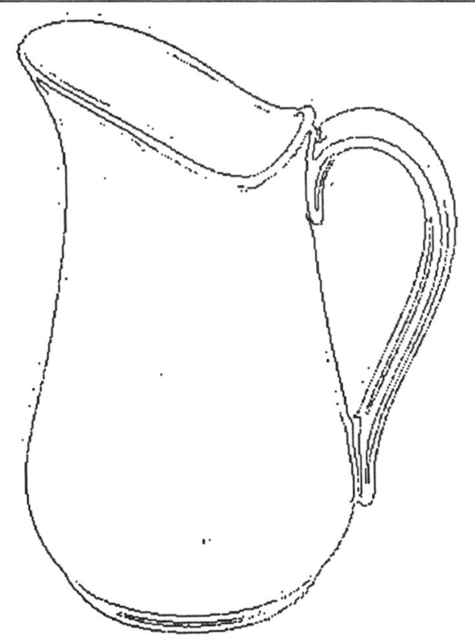

Figure 3.2 Thresholded zero-crossings.

3.5.1. FILLING IN THE GAPS

The goal of segmentation is to break the object up into regions bounded by closed contours of zero-crossings. If the convolved $\nabla^2 G$ image is thought of as a continuous 2-D function, then the zero-crossings form a closed continuous curve, segmenting the image. Due to the discrete nature of the convolution, the zero-crossings do not always form closed curves. Typically, small pixel gaps will appear, preventing a closed contour chain of 8-connected zero-crossings. A two stage algorithm (Bridge Gaps, algorithm 3.2) is used to close these gaps and form closed contours of zero-crossings. The first stage is a modification of a procedure of Nevatia and Babu [42] used to find linear segments from edge contours. This procedure creates a predecessor-successor array (PS). A PS array is created by designating the 8-connected predecessor and successor neighbors for every directed zero-crossing edge element. Edge elements that are at the beginning (end) of the 8-connected chain are designated as having no predecessors (successors). Edge elements that branch off with either two predecessors or two successors are also marked. From this array, chains of 8-connected zero-crossings are created. The second stage is to take these chains and to link them into longer chains, bridging gaps if needed. The second stage is an iterative process where successively longer chains are built and more pixels are bridged depending upon the pixel distance to be bridged. Initially, pixel gaps up to a distance of $2\sqrt{2}$ are bridged, requiring only a single pixel to be added. This stage repeats until only gaps of two pixels are left, at which point the two pixel gaps are iteratively filled. The algorithm will continue until the maximum designated gap distance is reached. In practice, filling in more than two pixel gaps is ambiguous. If a small filter size is used for the initial convolution, the zero-crossings are usually dense enough to fill the gaps accurately. Once the gaps are filled, the region analysis can continue.

This algorithm succeeds in filling small pixel gaps. However, certain imaging conditions will cause gaps greater than 3 pixels to be created. While the Bridge Gaps algorithm can span larger distances than 3 pixels, its performance degrades noticeably. For the zero-crossings in figure 3.2 the output of the Bridge Gaps algorithm successfully filled small contour gaps but was unable to bridge the gap at the top left corner where the surface turns sharply. In this image and the image of the coffee mug, small gaps that remained after the Bridge Gaps algorithm were filled by hand. This part of the segmentation problem in vision remains unsolved. A possible approach is to use scale space techniques and to follow zero-crossings at many levels to fill the gaps.

3.5.2. REGION GROWING

Region growing begins with the zero-crossing image which is output by the Bridge Gaps algorithm. This is an image containing zero-crossings and added pixels from the Bridge Gaps algorithm. Region analysis will separate the image into regions bounded by closed contours and will then calculate measures for each region. The algorithm to create each region from a closed contour (Region Grower, algorithm 3.3) is a recursive growing operation on the image that tries to grow a pixel's 4-connected neighbors until a border is found. As the algorithm grows these pixels, it colors them homogeneously, thus defining a region.

The algorithm uses two image arrays. Initially, the two arrays are identical with the zero-crossing image. A seed pixel is used to start a

---

Input: Zero-crossing image from Edge Detector algorithm
Maximum_Gap is maximum number of pixels to bridge

Output: Image with added pixels to create closed contours.

1. Form a predecessor-successor array (PS) that denotes the 8-connected neighbors that are predecessors and successors of directed zero-crossing elements. Mark beginning and ending elements and elements with multiple predecessors or successors
2. Starting at all beginning, ending or branch elements, traverse the connected chain and save it.
3. Set N=1.
4. Compare beginning or end elements of the chains. If the gap is less than N pixels, bridge the gap by adding the pixels and merging the chains.
5. Repeat step 4 until no N pixel gaps remain.
6. N=N+1.
7. IF N < Maximum_Gap goto step 4 else write out the image with added pixels.

Algorithm 3.2. Bridge Gaps.

---

growing operation that recursively grows the 4-connected neighbors of every pixel that is not an edge element. Each pixel that is grown is marked in the second array as visited with a particular color. When the recursive growing finally fails, all 4-connected pixels are colored homogeneously in the second array. By searching through the second array for an uncolored, non-edge pixel, we generate a new seed pixel and continue the operation, coloring grown pixels with a new color. This continues until all pixels in the second image are either colored or edge elements. The algorithm then examines each edge element in the colored image. The 8-connected neighbors of each edge element are compared and if they are all the same color or other edge elements, then this edge pixel is termed an *isolated* pixel, completely contained by a homogeneously colored region. Isolated pixels are colored

```
Image I1 and I2 are identical zero-crossing image arrays,
with pixels of value 0 or value 1 where an edge exists.
Output is homogeneously colored regions in image I2.

region_grower()
{   reg_color=2;
    FOR ( i=0; i<PICSIZE; i=i+1 ) {
       FOR ( j=0; j<PICSIZE; j<j+1 ) {
         IF (  I1[i][j] == 0 and I2[i][j] == 0 ) {
           grow(i,j,reg_color); /* non edge pixel, not visited */
         } /* end IF */
         reg_color = reg_color + 1;
       } /* end FOR */
    } /* end FOR */
    FOR ( i=0; i<PICSIZE; i=i+1 ) { /* remove isolated pixels */
       FOR ( j=0; j<PICSIZE; j<j+1 ) {
         IF ( I1[i][j] == 1 ) { /* is it an edge element? */
            homog(i,j); /* see  if the edge is isolated */
         } /* end IF */
       } /* end FOR */
    } /* end FOR */
} /* end region_grower */

grow (i,j,color)    /* grows 4-connected neighbors */
{  FOR ( k= -1; k<= 1; k=k+2 ) {
      FOR(m=0; m<2; m=m+1 ) {
        p=k ; q=0;
        IF ( I1[i+p][j+q] == 0 and I2[i+p][j+q] == 0 ) {
          I2[i+p][j+q] = color; /* mark as visited */
          grow(i+k,j,color);
        p=0 ; q=k;
      } /* end FOR */
   } /* end FOR */
} /* end grow */
homog (i,j)  /* colors isolated edges */
{  IF ( all non-edge 8-neighbors of I2[i][j] are color K ) {
        I2[i][j] = K;
   } /* end IF */
} /* end homog */
```

Algorithm 3.3. Region Grower.

Input to this algorithm is the colored region array computed by algorithm 3.3. Output is chains of pixels bordering a homogeneous region. Define the 8-connected neighbors of a pixel as:

```
          3 | 2 | 1
          -------------
          4 |   | 0
          -------------
          5 | 6 | 7
start = pixel in region with edge as 4 neighbor
new = start;  /* beginning edge element in chain */
first = TRUE; /* first time through switch */
s = 6;  /*  neighbor search direction  */
WHILE ( ( start != new )  or ( first ) ) { /* not closed yet */
  found = FALSE; /* flag for new contour pixel */
  cycles = 0; /* if 3 cycles: single pixel region */
  WHILE ( found = FALSE and cycles < 3 ) {
    cycles = cycles + 1;
    IF ( (s-1 mod 8) neighbor in region R ) {
      s = ( s - 2) mod 8;
      found = TRUE;
    } else {
      new = (s-1 mod 8) neighbor , add new to chain;
      first = FALSE;
      IF ( s neighbor in region ) {
        found = TRUE;
      } else {
        new = s neighbor, add new to chain;
        first = FALSE;
        IF ( (s+1) mod 8 neighbor in region ) {
          found = TRUE;
        } else {
          new = (s+1 mod 8) neighbor , add new to chain;
          first = FALSE;
          s = (s+2) mod 8;
        } /* end IF */
      } /* end IF */
    } /* end IF */
  } /* end WHILE */
} /* end WHILE */
```

Algorithm 3.4. Contour Tracer.

---

by their containing region's color.

The final part of the region growing is to output a chain of pixels that determines the closed contour of the region. This algorithm (Contour Tracer, algorithm 3.4) outputs a chain of 8-connected pixels that is the boundary contour of each region. The algorithm is a modified version of Pavlidis' contour tracer [50]. In Pavlidis' algorithm, a connected closed contour of a homogeneous region R is found by walking along the extremities of the region and recording the members of R who have neighbors not in R. The algorithm begins by finding a member of R with a neighbor not in R and always "walks to the right" finding 8-connected neighbors that are in region R with neighbors outside the region. This algorithm will output a chain of pixels that includes only members of the set R. What is desired instead is the chain of pixels not in R, but that have neighbors in R. In terms of the region picture from algorithm 3.3, we want the chain of edge pixels that separate regions, not the set of region points adjacent to the edges. The difference is important as the locations along the contours will be used for stereo matching.

The output of the Region Grower algorithm is an array of colored regions separated by closed contour edge chains. Figure 3.3 shows the closed contours formed for the pitcher by the Region Grower and Contour Tracer algorithms.

3.5.3. REGION ANALYSIS

These regions need to be further analyzed so that we may compute their centroids, average gray value and 2-D area. The centroid is used to find a beginning exploration point on the region and the area measure is used to order the regions for exploration. An important piece of information about these regions is their adjacency. From the region image, we can compute a region adjacency graph as defined by Pavlidis [50]. This is a graph that contains nodes which are colored regions and arcs between regions if they are adjacent. These adjacency relations will be used later in matching against the model data base. The adjacent regions are found by examining contour pixels that separate regions and by looking at the colors of their 8-connected neighbors.

Figure 3.3. Closed contours from Region Grower and Contour Tracer algorithms.

3.6. SUMMARY

The 2-D vision processing routines create bounded regions that can be used by the stereo matcher and tactile exploration algorithms. These algorithms create regions of larger interest moving away from pixel based point properties to token based contours and regions. As is the case in all vision processing, the tokens are artifacts of the lighting, reflectance and geometry of the surfaces imaged. The stereo algorithms in chapter 4 and the tactile exploration discussed in chapter 5 are intended to further classify these regions as surfaces, holes or cavities.

CHAPTER 4

3-D VISION

4.1. INTRODUCTION

Machine vision research has centered on the problem of obtaining depth and surface orientation from an image, creating what has been called by some authors the "2½-D" sketch [39]. Currently, there are several sensing systems that can derive depth or surface orientation from a scene. Among these are laser range finders[1,38,64], photometric stereo [29] and binocular stereo [5]. Laser imaging is potentially hazardous and has difficulty with shiny metal reflective surfaces. At present, it is a more expensive depth sensing technology than the other methods mentioned above. Photometric stereo puts great demands on the illumination in the scene and on properly understanding the reflectance properties of the objects to be viewed. We have chosen to use binocular stereo in this work because it has the advantage of low cost and the ability to perform over a wide range of illuminations and object domains. It is also a well understood and simple ranging method, which motivates its use in a generalized robotics environment where many different task and object domains may be in effect. Used as a single robotics sensing system, however, stereo has clear deficiencies. If there is a lack of detail on the object, only sparse measurements are possible. If too much detail is present, the matching process can easily become confused. Detail also causes a marked degradation in performance as the potential match space increases. This chapter discusses the stereo matching process used and analyzes the the ability of stereo to determine depth in our task and object domain.

4.2. COMPUTATIONAL STEREO

Stereo has been used in a variety of applications. A large body of work in stereo has centered on aerial photogrammetry, trying to determine object structure and depth from aerial images. Recently, interest in stereo for robotics has increased as the underlying visual processes in humans have been revealed. Barnard and Fischler [5] have broken down the computational stereo problem into a number of separate steps that are needed to generate depth representations from images. This chapter follows their paradigm and explains each step in the process in detail. The steps in the stereo process are:

- Image acquisition.
- Camera modeling.
- Camera calibration.
- Feature acquisition.
- Image matching.
- Depth determination.
- Interpolation.

4.3. IMAGE ACQUISITION

The camera system used to acquire the images is described in section 3.2. An important component of image acquisition is the domain of interest. In this research, the domain consists of smoothly curved objects with large surfaces, cavities and holes. The objects are not textured and are homogeneous in color, presenting a uniform albedo. The smooth nature of the objects and lack of textural detail are natural impediments to stereo matching, since these objects yield few match points.

4.4. CAMERA MODELING

In order to compute depth from stereo, a suitable camera model and camera parameters must be understood. Figure 3.1 shows the cameras used in this research. The two cameras are mounted with their focal points 12.7 cm. apart, defining what is known as the *stereo baseline.* The objects to be imaged are at a distance of 4 feet. In trying to find a correspondence between an event in one image and its counterpart, a large search problem exists. For an image of size $N \times N$ pixels, each pixel event in one image has potentially N^2 possible matches in the other image.. A simple and effective way to constrain this is to limit the search along epipolar lines. Epipolar lines (figure 4.1) are defined as the lines in each camera's focal plane caused by the intersection of the focal planes and the epipolar plane, which is a plane formed by a point P in the scene to be imaged and the two focal points of the cameras. A pixel event in one camera can limit its search for the corresponding event in the other camera to searching along the epipolar line in the corresponding camera. This effectively makes the search for an event $O(N)$ rather than $O(N^2)$. In a digital system, an effective approach is to register the cameras so that the epipolar lines correspond to the scan lines in the images. The procedure for registering the cameras is to take a test pattern of black circles and calculate the center of gravity of each circle in each image. The centers are then compared and adjustments made to have the centers correspond. The procedure produces correspondence of the centers of gravity to within 0.5 pixels across scan lines. This is a painstaking procedure that is extremely critical to the success of the stereo algorithms. The procedure is compounded by the additional camera parameters of focus and zoom which must also be adjusted for spatial coherence of the images.

4.5. CAMERA CALIBRATION

In order to determine depth, a transformation between the camera image coordinates and the 3-D world coordinate system being used is needed. This can be done in a number of ways. One method is to discover

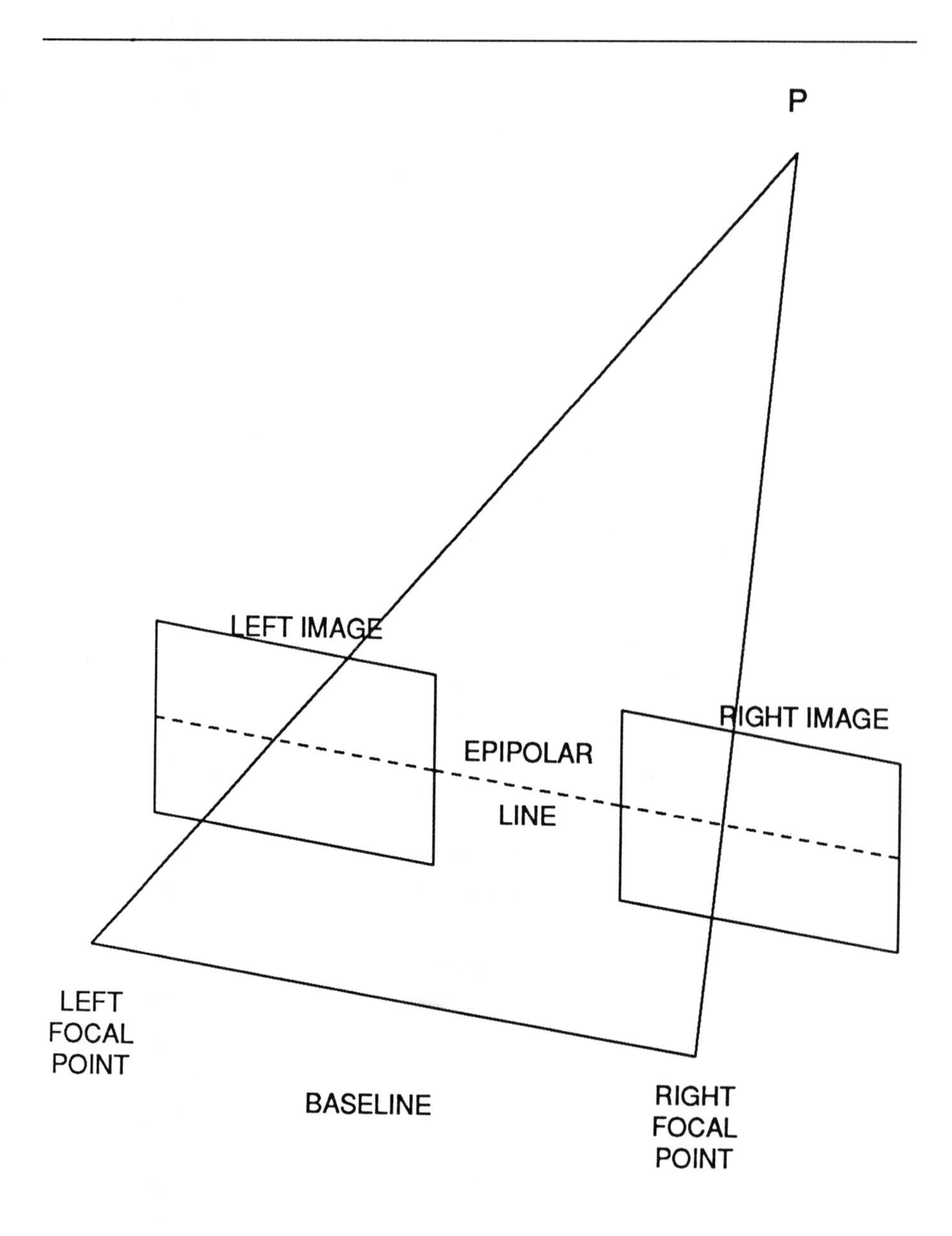

Figure 4.1. Epipolar lines.

the actual camera model parameters that relate the two coordinate systems [19]. Another method is to experimentally obtain a calibration transform from a series of known data points in the scene and the image [63]. This latter method is simpler and well suited to our problem. We can define a point in homogeneous 3-D world coordinates as:

$$\left[X , Y , Z , W \right] \tag{4.1}$$

and a homogeneous point in 2-D image space as:

$$\left[x , y , w \right] \tag{4.2}$$

The transformation matrix that relates these two coordinate systems is:

$$\left[X , Y , Z , 1 \right] \begin{bmatrix} T_{11} & T_{12} & T_{13} \\ T_{21} & T_{22} & T_{23} \\ T_{31} & T_{32} & T_{33} \\ T_{41} & T_{42} & T_{43} \end{bmatrix} = \left[x , y , w \right] = w \left[U , V , 1 \right] \tag{4.3}$$

Here we have arbitrarily set the homogeneous scaling factor W=1. If we multiply out these matrix equations, we get:

$$T_{11}X + T_{21}Y + T_{31}Z + T_{41} = w\ U \tag{4.4}$$

$$T_{12}X + T_{22}Y + T_{32}Z + T_{42} = w\ V \tag{4.5}$$

$$T_{13}X + T_{23}Y + T_{33}Z + T_{43} = w \tag{4.6}$$

If we substitute the value of w in (4.6) in (4.5) and (4.4) we get two new equations:

$$(T_{11}-T_{13}U)X + (T_{21}-T_{23}U)Y + (T_{31}-T_{33}U)Z + (T_{41}-T_{43}U) = 0 \quad (4.7)$$

$$(T_{12}-T_{13}V)X + (T_{22}-T_{23}V)Y + (T_{32}-T_{33}V)Z + (T_{42}-T_{43}V) = 0 \quad (4.8)$$

If we know a point (X, Y, Z) in 3-D world coordinate space and its corresponding image coordinates (U, V) then we can view this as a series of 2 equations in 12 unknown transform parameters T_{ij}. Since we get 2 equations per pair of world and image points we need a minimum of 6 pairs of world and image points to calculate the matrix. In practice, due to errors in the imaging system we will want to use an overdetermined system and perform a least square fit of the data. The technique used in solving an over-determined system of equations

$$A X = B \quad (4.9)$$

is to calculate the pseudo-inverse matrix and solve for X:

$$X = (A^T A)^{-1} A^T B \quad (4.10)$$

This method requires a way of determining the 3-D world points and the corresponding 2-D image points. The technique described here is due to Izaguirre, Pu and Summers [32]. The PUMA manipulator contains an embedded world coordinate system that is used to position the robot and is fixed to the robot's base. An LED is mounted on the end effector at a known position relative to the robot coordinate frame. The calibration procedure then moves the arm to one of a number of predetermined points in the camera's field of view. The LED is imaged in a dark room and the center of gravity of the LED impulse function in the image is computed, yielding sub-pixel image-space coordinates of the known 3-D world coordinates. Experimentation showed 40 points yielded low errors and a subsequent increase in the number of calibration points did not improve the accuracy. The errors in calibration were determined by substituting the calculated transformation parameters T_{ij} for each camera into equations 4.7 and 4.8 along with the known image coordinates in each camera and solving for X, Y and Z. This also is an overdetermined system of 4 equations in 3 unknowns that is solved by a least square fit. The transformation of each 2-D image point into 3-D is a line, and we are trying to find the intersection point in 3-D of the two lines emanating from the cameras. Due to imaging errors, these lines are usually skew, and the intersection point is the midpoint of the common perpendicular to these lines. The errors in position from the known 3-D robot positions were then computed. The largest error was in the X direction which relates directly to depth since the camera centers were generally aligned along this axis. The results for a typical calibration sequence of 40 points are in table 4.1.

4.6. FEATURE ACQUISITION

The correspondence problem for stereo is helped by isolating physical events in each image that correspond to the same location in space. Edges found by derivative-based operators are good candidates for features. They suffer from the point nature of the data which necessarily introduces small errors in the correspondence process. Researchers have sought to find larger groupings of pixels (tokens) in an attempt to lessen the effects of a single pixel error. Various tokens have been used. From edge detection algorithms lines and arcs have been isolated to try to match larger groupings of pixels with more accuracy. Gray level analysis has also tried to group regions of pixels showing similar gray level properties such as variance measures.

The features used in this work are the edge elements determined by the Edge Detector algorithm (algorithm 3.1). These image features correspond to the physical effects of geometry, lighting and reflectance. These edge elements are localized to subpixels and contain both magnitude and orientation information. A key element of the algorithm is the establishment of a threshold value in magnitude for a zero-crossing. Noise points which can cause problems for a stereo matcher are thinned out by this process. An important point to note is that this approach can err on the conservative side and still be successful. Most stereo systems have only vision to use; therefore decreasing the data gives rise to problems of sparseness. The approach being followed here is that relatively sparse (and correspondingly more accurate) visual data is supplemented with active tactile

CALIBRATION ERRORS, mm					
X	Y	Z	X	Y	Z
-1.108927	-0.058305	-0.301625	-0.478541	-0.073906	-0.123573
1.149862	0.081078	0.345332	0.514949	-0.017872	0.088030
0.812104	-0.026641	0.307286	-0.961732	-0.200877	-0.363145
-0.248319	-0.174759	-0.058981	-0.698619	0.105659	-0.280506
0.166807	0.050148	0.109394	-0.189775	0.187463	-0.093896
-0.902715	-0.051679	-0.375009	-1.037003	-0.049500	-0.363201
0.627194	0.010328	0.280518	0.543729	0.056435	0.168897
-1.137701	-0.024076	-0.377457	1.347597	-0.024106	0.492834
-0.148735	0.092887	-0.070734	0.858068	0.004397	0.341967
1.095666	-0.018231	0.511182	-0.089491	0.164923	-0.012673
-0.594617	0.003411	-0.310890	-0.737822	0.069448	-0.296553
0.583711	-0.013140	0.159542	0.334448	-0.014276	0.076215
-0.216956	-0.043942	-0.160494	-0.363198	-0.060250	-0.177463
0.409463	-0.117902	0.151622	0.653874	0.095679	0.340519
0.995076	0.110545	0.517877	0.106311	0.080782	0.110672
0.024526	0.135637	-0.048202	-0.281059	-0.074397	-0.055166
0.392691	-0.052172	0.182265	-0.426613	-0.220015	-0.197387
-0.027173	-0.003933	-0.007781	1.031748	-0.052976	0.323383
-0.177226	-0.141485	-0.074718	-0.443750	0.023856	-0.094611
-1.239145	-0.017965	-0.623783	-0.010544	0.155523	0.037553

Table 4.1. Calibration errors. The X axis measures depth from the the cameras.

exploration. Low confidence features are not used, nor are they needed in this approach.

A further thinning algorithm is used to make the matcher more accurate. The stereo matcher is only interested in matches along the closed contours of regions. All isolated edge pixels determined by the Region Grower algorithm (algorithm 3.3) are excluded from consideration by the matcher. This will greatly decrease the number of false matches seen by the matcher.

4.7. IMAGE MATCHING

This is the most difficult part of the stereo process. The image matcher used was originally developed by Smitley [60] for use on aerial images. It has been modified for the task domain of robotic object recognition. Given a set of features from each image, how do we match them? The initial matching criteria for two zero-crossing elements to match is:

- The zero-crossings must be on the same scan line.
- The zero-crossings must have a similar orientation.
- The zero-crossings must have the same contrast sign.

The initial constraint that helps here is the epipolar one: only features (zero-crossing edge elements) on corresponding scan lines are matched. This is not a strong enough constraint as there may be many edge elements in each scan line. The zero-crossings themselves provide us not only locality of the features but also magnitude and orientation information. If two edges match then their orientations should be similar in each image. The similarity measure used is 30°. A further requirement is that the contrast change across the edge be the same. Intuitively, this means that a black to white edge should match with a black to white edge in the other image, and vice versa for white to black edges. Requiring the edge magnitudes to correspond within a tolerance level does not prove to be helpful, although it is appealing to try to match edges by their "strength".

Many edges in the scan line can satisfy the weak criteria for selecting matches above. What is needed is a metric to measure the match after this initial matching stage so the matches may be ordered probabilistically. To establish a metric, a correlation is performed about windows centered on the matched pixels in each image. The output of the correlation is a metric of the degree to which the areas surrounding the matched pixels agree. High confidence levels (above 95%) are used, allowing only those matches that are robust to survive. Determining the size of the correlation window is an important part of the matching process. A small window will not include enough detail to disambiguate potential matches and a large window may drown out the effects of small local disambiguating features, at the cost of greatly increased processing time. A reasonable choice for this window size can be made by relating its size to the edge detector parameter w, defined in equation 3.2. The window over which correlation proceeds should be proportional to the density of the zero-crossings found. For a filter of size w, a window of size $2w \times 2w$ was used.

4.8. DEPTH DETERMINATION

Stereo algorithms are only as good as their ability to create dense and accurate depth maps. Unfortunately, most stereo algorithms have a number of serious problems which preclude them from reliably creating these depth maps. One of the more serious problems is that of homogeneous areas within images that are lacking in detail. No edge features are present in these regions and matching algorithms have no basis for a match. On the other hand, too much detail will confuse a matcher and cause false recognition. This is especially true with periodic textures on surfaces that leave little basis for local discrimination.

Another serious problem related to the practical implementation of stereo algorithms for depth determination is the inability of stereo to match edges whose orientation approaches horizontal. As edges become horizontal, localization of feature matches becomes ambiguous as is shown in figure 4.2. If we have a series of horizontally oriented zero-crossings in both images, then it is not at all clear how to match these points; they all satisfy the criteria of orientation and sign and within the correlation window have equally probable confidence levels. Experimentation has shown that as zero-crossing orientations approach 90° from vertical, the accuracy of the matches degrades seriously. Figure 4.3 shows the left and right closed contours of a coffee mug and the resulting correctly matched zero-crossings revealing the lack of horizontal match data. The matches were made with zero-crossings up to 70° from vertical; above 70°, the matches are unreliable.

A further implementation problem for determining depth from stereo is incorrect camera registration. An incorrect scan line registration of only one scan line can cause large errors. Figure 4.4 demonstrates this error. The two digital images of a curve are misaligned by one scan line, yet the resulting change in disparity is 4 pixels. This can translate to 16 mm in depth for the cameras being used. With vertically aligned edges the effect is minimized since the disparity values will be similar. As a digital curve approaches horizontal, the disparity values can change over a large range, causing "correct" matches but incorrect depth values.

Figure 4.2. Ambiguity of horizontal matches. Pixel A can match with B, C, D or E. The correlation windows will be identical in these regions.

Finally, stereo algorithms are subject to the errors caused by the discrete nature of a digital image. Solina [61] has analyzed the quantization errors due to stereo for the cameras used in this research. Figure 4.5 is taken from this work and graphically shows the error in location of two matched pixels. Any point within the diamond shape region will map to the same two pixels in the images. The error is a function of depth and increases as the distance from the camera increases. For the camera model used in this work, a one pixel error in disparity causes a change in absolute depth of approximately 4mm. By using subpixel accuracy, this error is reduced to 2mm.

4.9. INTERPOLATION

The last step in stereo vision processing is to interpolate the depth points calculated from stereo, resulting in the creation of a 2½-D sketch of the imaged surface. It is obvious from figure 4.4 that the data is too sparse to accurately interpolate a surface. Further, some of the regions are not surfaces but holes and cavities. If the system were relying on stereo vision alone, this would be another serious drawback to understanding the object's

Figure 4.3. Closed contours and stereo matches up to 65°.

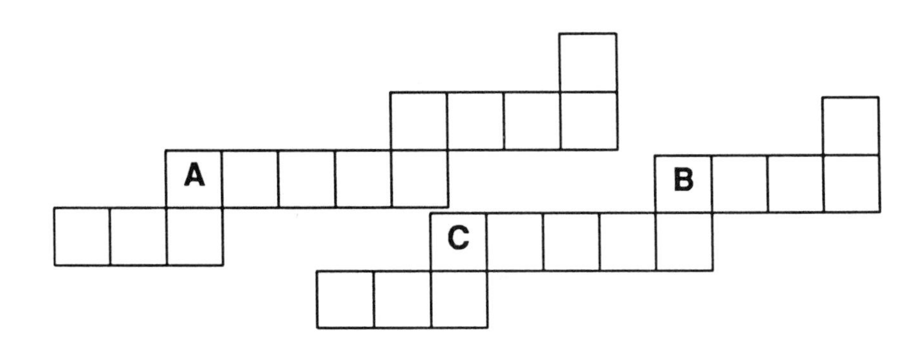

Figure 4.4. The digital curves are incorrectly registered by one scan line. Pixel A matches with B and not C, causing an error in disparity of 4 pixels.

structure. However, the tactile algorithms can fill in nicely what stereo cannot process. Multiple sensing allows a system to rely on each sensor for the data it can provide efficiently and accurately, rather than being dependent on a single modality. The intent of this work is to use those parts of vision systems that work well and not to try to have vision alone understand the scene. In the context of recognizing smooth objects without texture, stereo will be able to efficiently compute a sparse depth representation on the object's contour. This sparse contour data can be used to guide the active tactile exploration to fill out the surface and feature descriptions of the object to be recognized.

4.10. SUMMARY

While stereo appears to be a well understood visual process; its practical implementation in certain object domains leaves much to be desired; it is not robust enough to build dense surface descriptions for recognition purposes. However, the sparse matches provided by the stereo algorithms are reliable because they are based on contour tokens as opposed to pixels. High confidence levels are established for the matches in order to reduce

Figure 4.5. Stereo error. Points inside the diamonds have the same digital image coordinates.

error. This allows us to proceed to the next level of sensing with confidence, having sparse but accurate regions identified that can be used for further sensing. Attempts to drive the vision modules beyond this capability will invariably lead to a potentially serious error. The key idea is that *less is more* in the case of multiple sensing. We do not have to rely on this single modality for all our sensory inputs, only those it can reliably produce. The sparse and conservative matches produced are sufficient to allow tactile sensing to further explore the regions in space. Chapter 5 describes the nature of the tactile exploration algorithms and chapter 6 describes the integration of these two modalities.

CHAPTER 5

TACTILE SENSING

5.1. INTRODUCTION

Tactile sensing has for the most part been ignored in favor of other kinds of robotic sensing, particularly vision. The ability that humans have to infer 3-D shape and structure from projected 2-D images has led most researchers to try and emulate this human information processing ability. However, the complicated interaction and coupling of surface reflectance, lighting and occlusion yield intensity arrays that machines cannot understand well. The approach taken here is that for tasks such as object recognition, vision sensing is not enough. What is needed is more sensory information that can supplement the sparse and sometimes confusing visual data. In this work, the additional sensory data is supplied by a tactile sensor that is actively controlled and is used in an exploratory manner.

This chapter traces the development of tactile sensing in robotics environments with particular emphasis on the design and use of these sensors. It then describes the tactile sensor being used in this research. Lastly, the active tactile exploration algorithms that move the robotic arm using sensory feedback are described in detail.

5.2. CAPABILITIES OF TACTILE SENSORS

While vision remains the primary sensing modality in robotics, interest in tactile sensing is increasing. Harmon [24] has surveyed researchers in the field of robotics and reports that 90% of those surveyed viewed tactile sensing as an essential concomitant of vision. A major reason for this was the inability of vision systems to deal effectively with occlusion, uncontrolled illumination and reflectance properties. These researchers felt that the present state of 3-D scene analysis from vision was "pre stone age." They felt that tactile sensing systems would be part of an overall sensing environment that included many different kinds of sensors. Tactile sensing was felt to be important for recognition tasks, assembly and parts fitting work and inspection tasks. Tasks that call for close tolerances or low absolute error can benefit from a tactile approach. It seems clear that in a robotics environment intelligent touch is useful.

Tactile sensors vary in their ability to sense a surface. At the lowest level, simple binary contact sensors such as microswitches report 3-D coordinates of a contact point. The next level of sensor reports gray values that are proportional to the force or displacement on the sensor. The most capable of these sensors can also sense surface orientation, returning a surface normal vector. Useful properties that remain unexploited are temperature and hardness sensing. The geometries of these sensors vary from a single sensor to planar arrays of sensors to finger-like arrays covered with sensors. Much of the research in tactile sensing has centered on the transduction technology. A number of technologies including microswitches, strain gauges, piezoelectric materials and conductive elastomers have been utilized. For a thorough review of these technologies see Harmon [23].

An early effort at pattern recognition with tactile sensors was the work of Kinoshita, Aida and Mori [35]. They utilized a five-fingered hand containing 22 binary sensors to discriminate among objects. Each object was grasped from a number of different vantage points and the resulting binary pattern recorded. A discriminating plane was calculated in the sensor space from these learning samples. Then, the object is grasped a number of times and its membership in the discrimination space is computed. This work was able to distinguish a square pillar from a cylinder at 90% reliability. A similar approach was used by Okada and Tsuchiya [45] who used an eleven degree of freedom three-fingered hand to grasp objects and form binary patterns with the hand's contact sensors.

Another example of tactile recognition was the work of Ozaki et al [49]. In this work objects were treated as containing parallel slices which were sensed by a special gripper. The gripper consisted of 7 contact surfaces with tactile sensors (one palmar segment and two three segment fingers) which were wrapped around an object's contour and reported the unit normal distribution along the contour. This distribution was then matched with a set of model distributions to try to discriminate shapes. The system would not work well with objects that could not easily be described as a series of slices.

Overton [48] described a tactile sensor organized in a rectangular array capable of yielding gray value information proportional to force exerted on each sensor in the array (each sensor is a *forcel*). Simple vision array operators were used to distinguish patterns of tools from static sensing. A similar effort was reported by Hillis [27] who used a very high spatial resolution tactile sensor to distinguish small objects (screws, clips, bolts, etc.). His approach also was to use traditional gray level processing techniques on the array values to find bumps and holes on the surface. He also implemented a measure of the ease with which an object could be rolled. Because the sensor was larger than the object, static sensing was used.

5.3. SUMMARY OF TACTILE SENSING

Tactile sensing is still in its infancy. The approaches so far have emphasized traditional pattern recognition paradigms on arrays of sensor data, similar to early machine vision work. Most sensing has been static in that the sensor is larger than the object and a single "touch" is used for recognition. Very little has been done on dynamic sensing and on integrating multiple "touch frames" into a single view of an object. Gray value processing to determine surface properties is also limited. It appears that a fruitful approach to tactile sensing may be to follow the human paradigm,

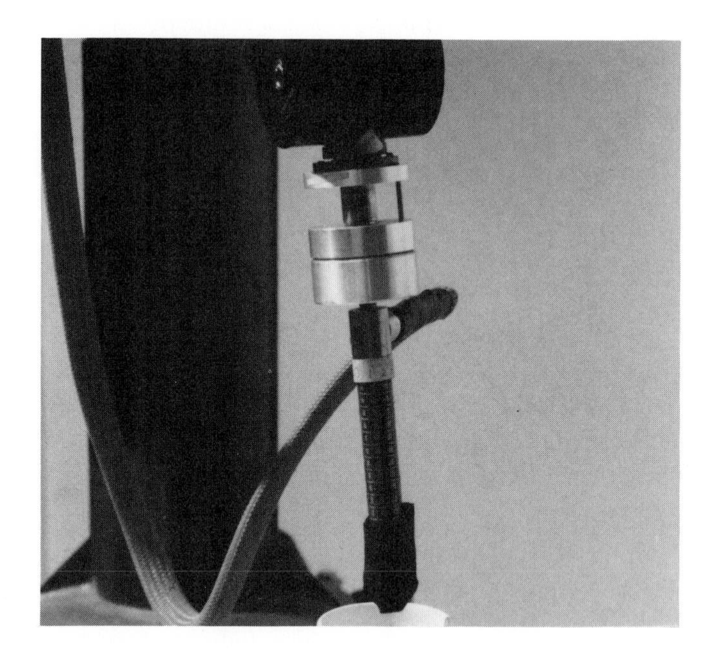

Figure 5.1. Tactile Sensor

identifying human tactile sensing properties to be used in machine tactile sensing (see Gordon [20] for an overview of human tactual perception).

Because tactile sensing is new and unexploited, major strides in many areas still need to be made. Among these is the need for more robust sensor design to increase spatial resolution, to eliminate nonlinearities and hysteresis and to increase dynamic range and bandwidth. In addition, intelligent control of sensors is needed at the software level as is the integration of these sensors into a multi-sensor environment. Solution of these problems will allow tactile sensing to be an important part of robotics systems, especially since it is potentially low in cost.

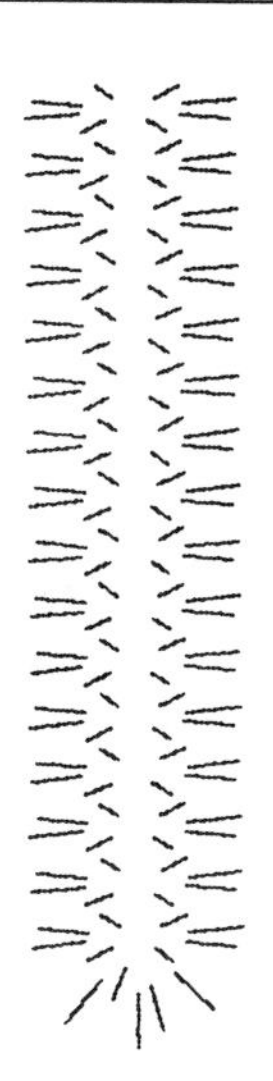

Figure 5.2. Surface normals on the tactile sensor.

5.4. EXPERIMENTAL TACTILE SENSOR

The experimental tactile sensor (figure 5.1) used in this research was developed at L.A.A.S in Toulouse, France. It consists of a rigid plastic core covered with 133 conducting surfaces. The geometry of the sensor is an octagonal cylinder of length 228 mm and radius 20 mm On each of the eight sides of the cylinder there are 16 equally spaced conducting surfaces. The tip of the sensor contains one conducting surface, and there are four other sensors located on alternate tapered sides leading to the tip. The tip sensing element is referred to as the *tip sensor,* the tapered sensors are referred to as the *taper sensors* and the sensors along each of the 8 vertical columns are referred to as the *side sensors.* Figure 5.2 shows the range of surface normal directions for each of the 133 sensing elements. The conducting surfaces are covered by a conductive elastomeric foam. The foam is produced in different widths from 2 mm to 4 mm which allows for a variation in compliance depending upon the task. There is a cable exiting from the top of the sensor that carries the reference signal and output wires from the sensors. This cable is connected to an analog to digital converter that outputs the readings on all sensors in an eight bit gray value. The entire array of sensors may be read in a few milliseconds. The digitized signal from the sensor A/D unit is fed into a Z-80 microprocessor that is responsible for the low level tactile processing.

The response characteristics of the sensor vary slightly over the 133 sensing elements. A representative sensing element is the tip element. Contact forces of 170 grams resulted in a gray value reading of 1, 453 grams yielded a gray value of 127 and 1100 grams saturated the sensor with a gray value of 255. The repeatability of the sensor can also be measured. Table 5.1 shows the X Y Z coordinates of contact reported by the sensor during a test of repeatedly moving the sensor onto and off of a rigid surface. The spatial resolution of the sensor is relatively poor. The side sensors are approximately 8 mm. apart and the tip sensors 7 mm. Thus, localization of signal can cause an error of up to 4mm.

The sensor is mounted on the end effector of a PUMA 560 manipulator [65]. This is a commercial six degree of freedom robotic manipulator. The tactile sensor is mounted with its long axis perpendicular to the mounting plate. This is called the *tool Z axis*. There is a mechanical overload protector in the mounting plate of the sensor which will allow the sensor to deflect, preventing it from being damaged by an accident, if a force greater than approximately 5 pounds is exerted on the sensor.

5.5. ORGANIZATION OF TACTILE PROCESSING

The organization of tactile sensing is on three distinct hardware and software levels (figure 5.3). The highest level consists of programs on a VAX host that provide high level control information about the regions in space that are to be explored with the sensor. Algorithms have been developed to explore the regions isolated from the vision processing and determine if they are surfaces, holes or cavities. Once a region is identified by tactile sensing, it can be further explored by tactile surface following algorithms that report contact points on surfaces and boundary contours of holes and cavities to the controlling host process. These contacts can be integrated with the 3-D contours from vision to build robust surface and feature descriptions. The intermediate level consists of programs written in VAL-II [66] that run on the PUMA and move the robotic arm based upon feedback from the tactile sensor. The intermediate level receives region exploration parameters via the

CONTACT COORDINATES, mm		
x	y	z
222.38	613.16	-286.97
222.47	613.16	-286.16
222.22	612.91	-286.81
222.41	613.16	-286.72
222.50	613.19	-286.56
222.38	613.25	-286.84
222.38	613.22	-287.13
222.47	613.03	-286.75
222.47	613.00	-286.47
222.50	613.19	-287.19
222.38	613.06	-287.00
222.38	613.09	-286.75
222.34	612.97	-286.75
222.41	613.06	-286.41
222.47	613.03	-286.91
222.38	613.25	-287.25
222.34	613.03	-286.63
222.22	612.91	-286.88
222.50	613.16	-286.63
222.22	612.94	-286.88
STATISTICS		
$\bar{x}$	$\bar{y}$	$\bar{z}$
222.39	613.09	-286.78
σ_x	σ_y	σ_z
0.09	0.11	0.27

Table 5.1. Repeatability of sensor contact.

VAL-II's host control mechanism which are used to invoke a surface exploration, hole exploration or cavity exploration procedure. These procedures use the feedback from the tactile sensor contacts to control arm motion along the exploration path determined by the high level host control. The intermediate level communicates with the low level sensor system via commands that set thresholds for contacts, requests contact interrupts and requests gray level outputs from arbitrary subsets of the sensor's elements. The low level system is implemented on a micro-processor that samples, digitizes, conditions and localizes the data coming from the tactile sensor, interrupting the

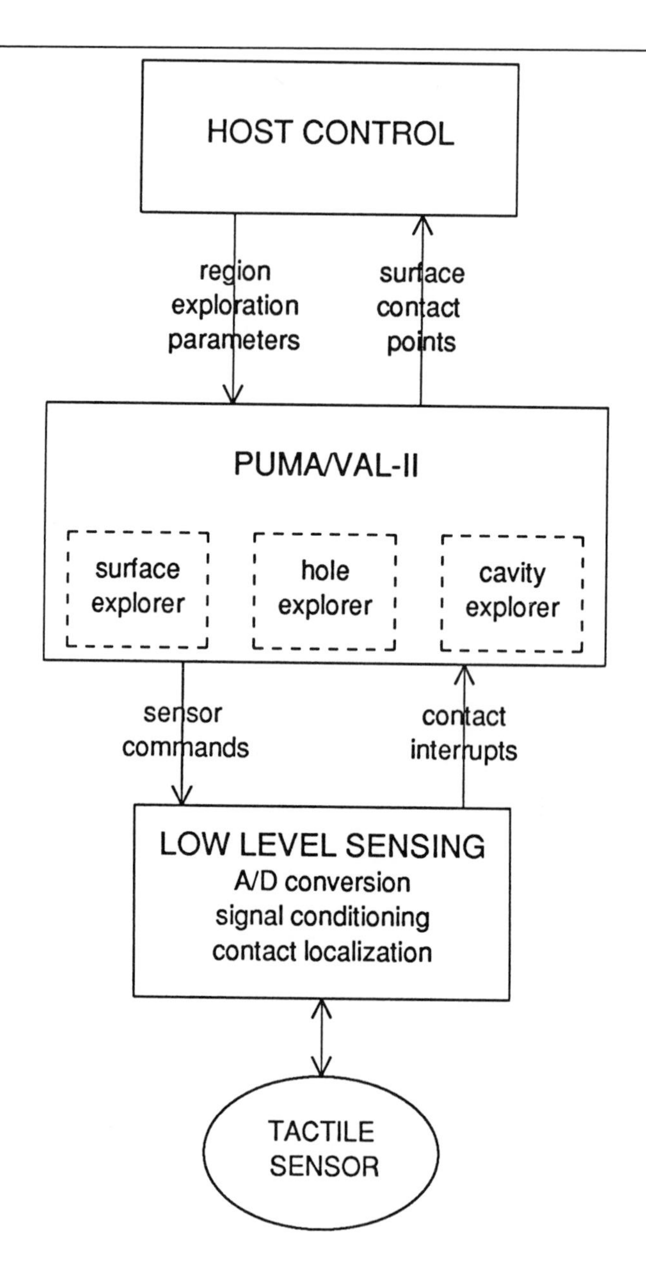

Figure 5.3. Tactile sensing system.

intermediate level if a contact of a certain nature occurs. The low and intermediate levels will be explained in the following sections. The high level includes integration of vision as well as tactile data and is explained in chapter 6.

5.6. LOW LEVEL TACTILE PROCESSING

The low level tactile processing is a series of Z-80 programs that condition and filter the data coming from the sensor. The low level routines on the Z-80 work in conjunction with the intermediate level tactile routines in the PUMA controller that move the PUMA arm with tactile sensory feedback.

The Z-80 executes a series of commands that are specified from the PUMA system. The Z-80 maintains an internal array of the 133 contact sites most recent gray value readings and has the ability to take new readings from the sensor. The low level routines that are performed on the Z-80 are explained below.

- *Set global threshold.* This function establishes a threshold gray value. Any contact that is below this value is ignored.
- *Set local threshold.* This allows a mask to be specified with varying thresholds for each sensor. The main function of this command is to normalize the signal response for all 133 sensors.
- *Snapshot.* This command causes the Z-80 to poll all 133 sensor sites in order and report back the gray values for each sensor. The command returns a list of 133 ordered pairs consisting of sensor number and gray value.
- *Sorted snapshot.* This is the same as the *snapshot* command except that the ordered pairs are presorted by gray value before being sent back to the PUMA system.
- *Return number of over threshold sensors.* This command causes the Z-80 to interrogate the sensors and report back the number of sensors over the global threshold gray value. This command is useful in comparing changes over time (moving on to or off of a surface) and in ignoring spurious responses.
- *Return N locations over threshold.* This command will return the N locations and gray values that are over the global threshold value,

sorted by decreasing gray value. If N=1, then this will return the location and gray value of the sensor with the maximum contact force applied. If N=133, then this command functions identically to *sorted snapshot.*

- *Guarded move.* This command will cause the Z-80 to continuously monitor each of the 133 sites on the sensor and report back (via an interrupt to the PUMA) the location and gray value of the sensor with the highest over-threshold gray value. This is the most useful primitive for surface following and movement of the arm with feedback.
- *Nearest neighbors.* This command asks for the values of the 4 nearest neighbors of a specific sensor. Once the maximum contact sensor is found, establishing contact values at neighboring sites will allow better localization of contact and determination of potential spurious signals.

5.7. INTERMEDIATE LEVEL TACTILE PROCESSING

The intermediate level tactile processing takes place in the VAL-II system of the PUMA. VAL-II [66] is a robot programming language developed for the PUMA series of robots. A particularly useful feature of VAL-II is host control. This allows another computer to act as a controlling node for the VAL-II system. Using host control, VAL-II commands can be issued on the host and transmitted over a serial link to the PUMA where they are then executed. All program I/O with the VAL-II system is sent to and from the host machine. The effect of this is to allow the high level control module of the object recognition system to directly call the VAL-II commands to perform arm movement with tactile feedback. This procedure has been simplified by a set of C language subroutines written by Alberto Izaguirre [31] that duplicate the VAL-II command set, allowing a C program on the host computer to use the VAL-II command set and move the robotic arm.

The PUMA has an embedded world coordinate system whose origin is centered at the intersection of joint 1 (waist rotation) and joint 2 (upper arm). A location in this space is specified in VAL-II as a 6-vector $[x, y, z, o, a, t]$, where x, y, z are the translational parameters and o, a, t are modified Euler angles used to determine orientation. The special location HERE returns the 6-vector that corresponds to the position and orientation of the end effector, measured at the center of the tool mounting surface on the wrist. VAL-II allows the designation of arbitrary coordinate frames by supplying a frame origin and two axes for the frame. This allows representing locations in the coordinate frame of the tactile sensor, once the frame that represents the sensor is defined. Each sensing element's position in space is then defined as a relative transform from the tool mounting plate, allowing computation of its absolute position in space. The orientation of each sensing element is also known, allowing computation of the surface normal at the contact site within the limits of the sensor's orientation resolution.

VAL-II has commands to allow asynchronous interrupts on 16 binary sensor I/O lines. If the low level tactile processing determines that an over threshold contact has occurred, an interrupt can be sent to the VAL-II system. This interrupt will cause the automatic invocation of an interrupt service routine which can communicate with the arm movement programs via a shared memory location, causing the movement of the arm to be modified based upon the position and orientation of the tactile contact.

The intermediate level tactile processing is characterized by the need to integrate the low level tactile sensor feedback with the coordinated movement of the arm. The arm needs to be used as an exploratory device. It is guided by high level knowledge about each region to be explored and the low level sensory feedback from surface contact.

5.7.1. EXPLORING REGIONS

The high level tactile processing will determine a region to explore by touch. Once a region is chosen to be explored, the intermediate level VAL-II exploration program is remotely executed by the high level using VAL-II host control. The Explore Region algorithm (algorithm 5.1) will establish if the region discovered by the vision algorithms is a surface, hole or cavity. The program needs as input an approach vector towards the region. The computation of this approach vector is important since it requires specifying a starting position in space to which the tactile sensor must be moved and also an orientation which represents the direction from which the sensor will approach the region. The orientation of the sensor is computed by calculating the least square plane P_{lsq} with unit normal $\mathbf{N}_{lsq}$ from the matched 3-D stereo points that form the contour of the region. $\mathbf{N}_{lsq}$ then becomes the approach vector for the sensor. The VAL-II routines will then orient the arm so that the tactile sensor's long axis (tool Z axis in the sensor frame) is

Algorithm to determine if a region is a surface, hole or cavity.
Inputs:
P_{lsq}: Equation of least square plane of 3-D points
of region's contour with unit normal $\mathbf{N}_{lsq}$.
C_{start}: Intersection point in 3-D between the line L_1
from the camera back projected through the 2-D
centroid of the region and P_{lsq}.
Outputs: Classification of the region as a surface,
hole or cavity.
BEGIN.
Build coordinate frame T1 with $\mathbf{N}_{lsq}$ as Z axis
and C_{start} as origin of frame T1.
T2 = frame T1 translated to workspace bounds along $\mathbf{N}_{lsq}$.
MOVE arm to T2. /* aligns sensor with plane normal */
Set global threshold for tactile sensor.
Set up guarded move interrupt.
DIST = 0. /* distance sensor tip has moved past P_{lsq} */
SENSOR_LEN= length of tactile sensor along its Z axis.
REPEAT
  MOVE along positive Z axis of frame T2 1 mm
  IF (sensor tip has moved beyond P_{lsq}) {
    DIST = distance between sensor tip and P_{lsq}
  }
UNTIL ((tip contact established) or
    (DIST > SENSOR_LEN)). /* contact a surface or hole found */
IF (tip contact established and DIST $< T_{cav}$) {
    Set CENTER = tip contact point.
    report "surface" to host.
} ELSE {
    IF (DIST >= T_{cav} and DIST < SENSOR_LEN) {
      CAVITY_DEPTH = DIST.
      CAVITY_BOTTOM=HERE.
      report "cavity" to host.
    } ELSE {
      HOLE_CENTER = HERE.
      report "hole" to host.
    } /* end IF */
} /* end IF */
END.

Algorithm 5.1. Explore Region.

aligned with $\mathbf{N}_{lsq}$. The starting point C_{start} to which the sensor is moved is calculated by intersecting plane P_{lsq} with the line L_1 formed by back projecting the region's 2-D centroid into the scene. C_{start} is then modified by translating it back along $\mathbf{N}_{lsq}$ so it is off any surface that might be in that region.

The arm is then moved along the tool Z axis until it contacts a surface or it moves beyond plane P_{lsq}, implying the presence of a hole or a cavity. If the sensor is able to travel its full length beyond P_{lsq} without contact, then a hole has been found. If it travels beyond a specified cavity threshold T_{cav} before contact, then it is a cavity.

5.7.2. SURFACE TRACING

Once the sensing routines have determined if the region is a surface or a hole or a cavity, the region must be further explored. If the region is a surface, then a bicubic surface patch must be built by integrating vision and touch. This procedure is explained in chapter 6. What is required of the intermediate level routines is to trace across the surface that has been discovered, reporting back points of contact along the way. These contact points on the surface are then integrated by the high level tactile processing into a surface patch describing the surface. The Trace Surface algorithm (algorithm 5.2) takes as input the point CENTER defined in algorithm 5.1, which is the 3-D point where contact with the surface was established. The trace routines trace out from this point to edges of the region. The high level routines choose 4 knot points on the region's boundary to serve as knot points for the bicubic surface patch. These knots create four boundary curves around the region. By tracing out from the CENTER point towards the midpoints of each of the regions four boundary curves, a new knot set and boundary curves are created. This procedure is explained in detail in chapter 6.

There are many paths between two points on a surface. The constraints that are used in determining the path to traverse in this algorithm establishes a weighted move vector in determining the next movement along the surface toward the goal point.

The movement vector $\mathbf{M}$ is determined by:

$$\mathbf{M} = \sum_{i=1}^{3} w_i \, \mathbf{G}_i \tag{5.1}$$

w_i are the weights for each of the vectors $\mathbf{G}_i$.

$\mathbf{G}_1$ is the unit vector in the direction of boundary curve midpoint.

$\mathbf{G}_2$ is the unit vector formed from the previous two contact points.

$\mathbf{G}_3$ is the unit vector that preserves equal parameterization.

$\mathbf{G}_1$ is needed to make progress towards the boundary edge. We will want to make progress towards the boundary at each movement step. However, with concave and convex surfaces, cycles can occur as the trace progresses. $\mathbf{G}_2$ is used to maintain a path's direction. Once we start moving in a certain direction we do not want to stray too far too fast from that path. This vector is an "inertia" vector helping the sensor stay on a steady course. $\mathbf{G}_3$ is needed to keep the parameterization of the surface patches uniform, and this vector moves the trace in the direction to preserve parameterization. This vector is the unit resultant of the vectors from the present contact point on the surface to the endpoints of the boundary curve that the trace is approaching.

The surface trace begins by contacting the surface, determining the surface normal from the contact sensor element, and backing off in the negative surface normal direction a short distance. **M** is then calculated and the arm is moved a short distance in that direction. The surface is then recontacted along the surface normal and the cycle repeats. The trace is ended by incurring one of two conditions. The first condition is determination of a surface discontinuity. The regions to be explored are smooth from the vision analysis since they are lacking in zero-crossings in their interior. If an edge discontinuity appears (as signaled by side sensor contact), the trace will end since it has reached a surface geometry change. The other condition to end the trace is when the surface contact points are within a threshold of the boundary stereo match curve. Thus the trace will end on occlusion edges or discontinuity edges.

```
Algorithm to perform 3-D surface tracing
with tactile sensory feedback.

Inputs: CENTER is starting point on surface
        N_lsq is normal to P_lsq
        D is small movement distance
        GOAL is goal point of trace

Output: Series of contact points on the surface

BEGIN
MOVE arm to CENTER.
ALIGN arm with tool Z axis along N_lsq.
set global threshold for tactile sensor.
surface_normal= tool Z axis.
REPEAT
  REPEAT
    MOVE along surface_normal D mm
  UNTIL ( surface contact established ).
  report contact position to host.
  calculate surface_normal from contact sensor orientation.
  MOVE along negative surface_normal D mm /* back off */
  calculate M. /* from equation 5.1 */
  MOVE in direction M D mm
UNTIL ( GOAL reached or contact by side sensors ).
END.
```

Algorithm 5.2. Trace Surface.

```
T1 is coordinate frame from Explore Region algorithm.
D1 is small movement distance.
D2 is threshold for moving without contact. If movement is
  longer than D2, we need to recontact the surface.

MOVE arm to frame T1.
MOVE arm along Z axis of T1 D1 mm /* Z axis is long axis */
REPEAT
  MOVE along X axis of T1 D1 mm /* perpendicular to Z */
UNTIL (side contact established at point P_start).
report coordinates of contact point P_start to host.
P = P_start.
REPEAT
  distance_moved = 0.
  N1 = calculated surface normal at P.
  N2 = projection of N1 onto XY plane of frame T1.
  N3 = N2 rotated 45 ° about tool Z axis.
  MOVE off surface along N3 D1 mm /* back off in tool XY plane */
  N4 = N3 rotated 90° about tool Z axis. /* approach in XY */
  REPEAT
    IF ( distance_moved < D2 mm ) {
        MOVE towards surface along N4 D1 mm
        distance_moved = distance_moved + D1.
    } else { /* gone too far without contact */
        MOVE along negative N1 D1 mm /* recontact surface */
    }
  UNTIL (side contact established at P)
  report coordinates of P to host.
UNTIL ( distance from P_start to P < D ).
```

Algorithm 5.3. Trace Hole/Cavity.

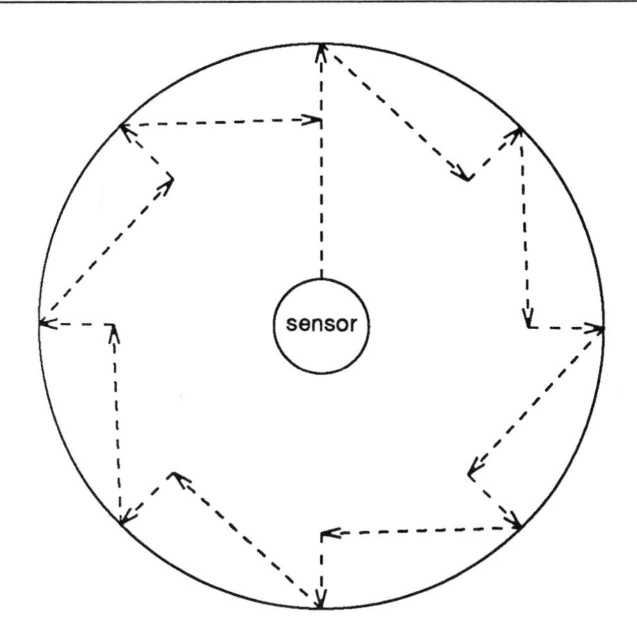

Figure 5.4. Movement of the tactile sensor inside a hole.

5.7.3. HOLE/CAVITY TRACING

If the Explore Region algorithm determines that the region is a hole or cavity, a different tactile tracing routine is used. In the case of a hole or cavity, we want to determine its cross-sectional area, moments and boundary. This can be done by moving the sensor around the hole or cavity's boundary and recording the contact points which are then sent to the high level routines for calculation of the properties mentioned above. The Trace Hole/Cavity algorithm (algorithm 5.3) begins by moving the sensor just beyond the least square plane P_{lsq} of a region's contour points, aligned with $\mathbf{N}_{lsq}$. It then proceeds to move in a direction perpendicular to $\mathbf{N}_{lsq}$ until it contacts a surface. Once the surface is contacted, the sensor moves in a sawtooth manner (figure 5.4) staying perpendicular to $\mathbf{N}_{lsq}$, alternately backing off and recontacting the surface, recording the contact points. The distance that the sensor travels between contacts is continually updated, and if it exceeds a threshold, the sensor will return to the surface along the last contact normal, recontacting the surface. This prevents the sensor from losing its way. When the sensor returns to the starting point, the trace is complete. The set of points recorded constitutes a boundary contour for the hole or cavity, which is then processed by the high level routines.

5.8. SUMMARY

The use of tactile sensing in robotics has been limited. Previous approaches have emphasized static sensing using traditional pattern recognition techniques. The approach taken here is to use active, dynamic sensing of surfaces and features to try to uncover the underlying 3-D structure of the object. The organization of tactile sensing is on three distinct hardware and software levels. The low level is a series of programs that condition and sample the data coming from the sensor. The intermediate level consists of programs that move the robotic arm based upon feedback from the tactile sensor. Algorithms exist to explore a region in space and determine if it is a surface, hole or cavity. Once a region is identified, it can be further explored by surface following algorithms that report contact points on surfaces and boundary contours of holes and cavities to a controlling host process. The high level knowledge needed to perform these tactile explorations is described in the next chapter.

CHAPTER 6

INTEGRATING VISION AND TOUCH

6.1. INTRODUCTION

The vision and tactile processing described in the previous chapters needs to be integrated to build 3-D descriptions of surfaces and features of objects that can be matched against the models in the model data base. The procedures described in this chapter use both sensing modalities, integrating the data from the sensors to build high level descriptions of what is seen and felt. The first integration procedure is a hierarchical procedure for building curvature-continuous composite surfaces from the vision and touch data. This procedure computes a Coons' patch representation which is the same primitive used in the model data base, facilitating matching between the sensed objects and the models. The second procedure is a method for creating smoothed boundaries of hole and cavity cross sections using both vision and touch which is accurate and also useful for matching.

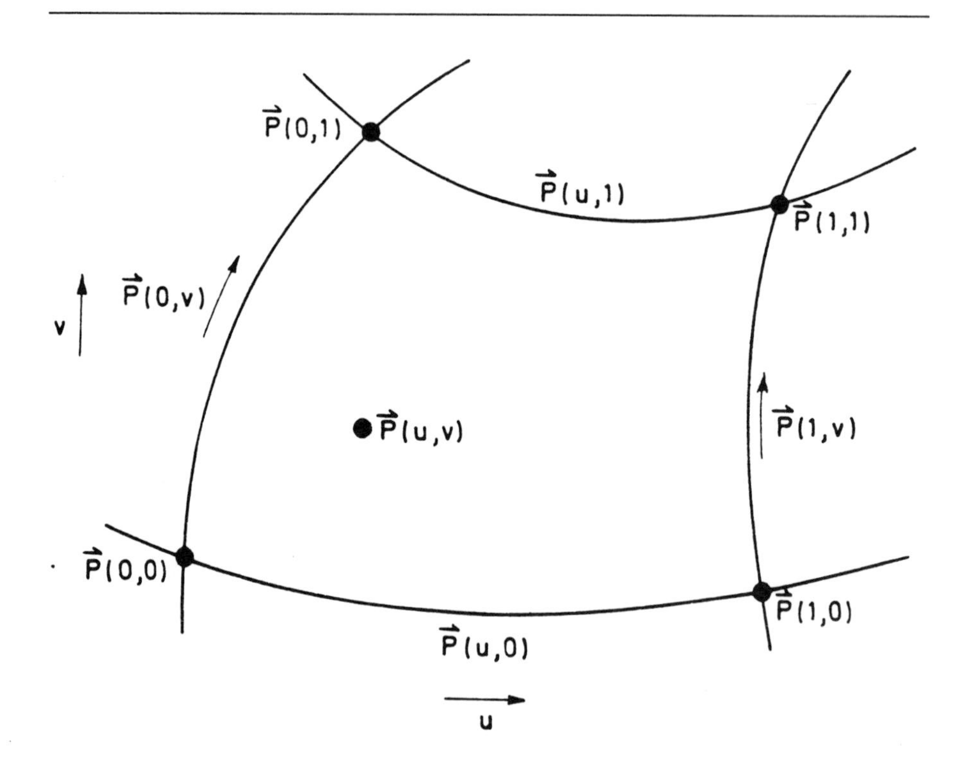

Figure 6.1. Parametric surface patch

6.2. COMPOSITE BICUBIC SURFACES

The parametric representation of a surface patch $\mathbf{P}(u,v)$ is shown in figure 6.1, where the parameters of the surface u,v range from 0 to 1. The patch has four knot points, $\mathbf{P}(0,0)$, $\mathbf{P}(0,1)$, $\mathbf{P}(1,0)$, $\mathbf{P}(1,1)$ which determine the endpoints of the boundary curves of the patch, $\mathbf{P}(0,v)$, $\mathbf{P}(u,0)$, $\mathbf{P}(1,v)$, $\mathbf{P}(u,1)$. A Coons' patch is fully determined by specifying the 4 knot points, the parametric derivatives at the knot points $\mathbf{P}_u$ and $\mathbf{P}_v$, and the parametric cross derivatives at the knot points $\mathbf{P}_{uv}$.

We can extend the notion of a single Coons' patch to a composite surface, containing a series of curvature-continuous patches defined on an arbitrary knot set. The information needed to create a series of curvature-continuous patches on a set of $M \times N$ data points $\mathbf{P}(u,v)$ defined on a rectangular parametric mesh is summarized in figure 6.2.

$\mathbf{P}_{uv}(0, N)$	$\mathbf{P}_{v}(0, N)$	$\mathbf{P}_{v}(1, N)$	– – –	$\mathbf{P}_{v}(M, N)$	$\mathbf{P}_{uv}(M, N)$
$\mathbf{P}_{u}(0, N)$	$\mathbf{P}(0, N)$	$\mathbf{P}(1, N)$	– – –	$\mathbf{P}(M, N)$	$\mathbf{P}_{u}(M, N)$
$\mathbf{P}_{u}(0, N-1)$	$\mathbf{P}(0, N-1)$	$\mathbf{P}(1, N-1)$	– – –	$\mathbf{P}(M, N-1)$	$\mathbf{P}_{u}(M, N-1)$
–	–	–	– – –	–	–
–	–	–	– – –	–	–
–	–	–	– – –	–	–
$\mathbf{P}_{u}(0,0)$	$\mathbf{P}(0,0)$	$\mathbf{P}(1,0)$	– – –	$\mathbf{P}(M,0)$	$\mathbf{P}_{u}(M,0)$
$\mathbf{P}_{uv}(0,0)$	$\mathbf{P}_{v}(0,0)$	$\mathbf{P}_{v}(1,0)$	– – –	$\mathbf{P}_{v}(M,0)$	$\mathbf{P}_{uv}(M,0)$

$\mathbf{P}(i,j)$ are the data points defined on the grid.

$\mathbf{P}_u(i,j)$ are the tangents in the u direction.

$\mathbf{P}_v(i,j)$ are the tangents in the v direction.

$\mathbf{P}_{uv}(i,j)$ are the cross derivatives or twist vectors.

Figure 6.2. Information needed to build a composite surface.

To build an interpolating composite surface, all that is needed besides the data points themselves is tangential and twist vector information at the boundaries of the mesh. The main task of the surface building algorithms is to compute the information in figure 6.2 using data from both vision and tactile sensors.

6.3. BUILDING LEVEL 0 SURFACES

Level 0 surfaces are surfaces comprised of a single surface patch. They are defined on a 2 x 2 rectangular knot set. The information needed besides the 4 knot points are the tangents in each of the parametric directions and the twist vectors at these knots. The choice of the knot points on the boundary of the surface is important. If these points are not chosen

---

Algorithm to choose 4 knot points of high curvature uniformly spaced on the contour.

Input: Digital 2-D closed curve C, consisting of points P_i. D is a minimum knot separation distance along the curve. L is a threshold for boundary curve length equality.

Output: Location on C of the 4 knot points.

1. For every point i on C, compute c_{ik} for a range of k in the neighborhood of P_i. Compute the maximum of these c_{ik} and store it as c_i also storing the range h where h is the neighborhood around pixel i where the curvature maximum occurred.

2. For every point on the contour, if $c_i \geq c_j$ for all j in neighborhood $\frac{h}{2}$ of P_i, then save this c_i as a local maximum.

3. Order the c_i determined in step 2 by cosine value.

4. Let the initial knot be P_j where j is the location on the curve where the largest curvature was seen.

5. Choose two more knots from the ordered list. If any one of the chosen curvature maximums is within a neighborhood D of an already chosen point, reject this point as a knot.

6. Choose the final knot from the ordered list such that the difference in lengths of the opposing boundary curves is less than L.

Algorithm 6.1. Choose Knot Points.

---

wisely, the resulting surface will be a poor approximation to the real surface. There are two considerations in choosing the knot points. The first is that the points should be chosen at points of high curvature on the boundary curve. If the parametric direction tangents coincide with the lines of curvature on the surface, then the twist vectors will be zero, which will allow a simple computation of the surface. The second consideration is that the knots need to be spaced uniformly in each of the parametric directions. Given a closed contour boundary of a region from the vision algorithms, we have to choose the four corner knots that will be used to create a level 0 surface. The algorithm that does this chooses these points according to curvature and parametric spacing. The Choose Knot Points algorithm (algorithm 6.1) for choosing points of high curvature on a contour is a modification of an algorithm originally proposed by Johnston and Rosenfeld [55]. Given a set of contour boundary points $\mathbf{P}_i$ the k vectors $\mathbf{A}_{ik}$ and $\mathbf{B}_{ik}$ at $\mathbf{P}_i$ are defined to be:

$$\mathbf{A}_{ik} = \mathbf{P}_i - \mathbf{P}_{i+k} \tag{6.1}$$
$$\mathbf{B}_{ik} = \mathbf{P}_i - \mathbf{P}_{i-k}$$

The k *cosine* at $\mathbf{P}_i$ is:

$$c_{ik} = \frac{\mathbf{A}_{ik} \cdot \mathbf{B}_{ik}}{|\mathbf{A}_{ik}|\,|\mathbf{B}_{ik}|} \tag{6.2}$$

In this definition, c_{ik} is the cosine of the angle formed between the k vectors $\mathbf{A}_{ik}$ and $\mathbf{B}_{ik}$. Accordingly, points of high curvature will have a cosine of +1 (zero angle between them) and points with no curvature will have a value of -1 (lying on a straight segment). The algorithm computes c_{ik} for a range of k in the vicinity of $\mathbf{P}_i$. It assigns a level h at each $\mathbf{P}_i$ where h is the value of k that maximizes c_{ik}. This yields a set of local maxima of curvatures that are then further thinned by retaining only those local maxima c_{ih} that are greater than or equal to any other local maxima within range $\frac{h}{2}$ of $\mathbf{P}_i$.

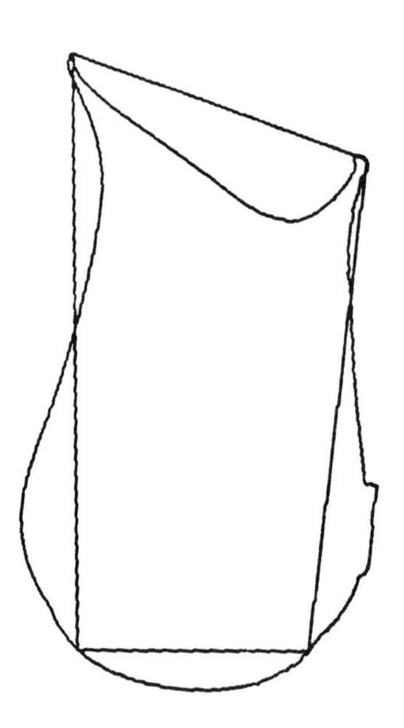

Figure 6.3. Knot points chosen on pitcher surface.

This algorithm yields the curvature values at each point of the contour. Starting with the maximum curvature value found, the four knots are successively chosen. Any point of high curvature that is within a distance D of an already chosen point is rejected to insure uniform spacing. The final knot creates a series of 4 boundary curves on the contour. An important requirement of this method is that boundary curves on opposite sides of the contour be approximately equal in length. Step 6 of the algorithm reflects this constraint. Figure 6.3 shows the closed contours developed from the image of a pitcher and the chosen knot points on the contour.

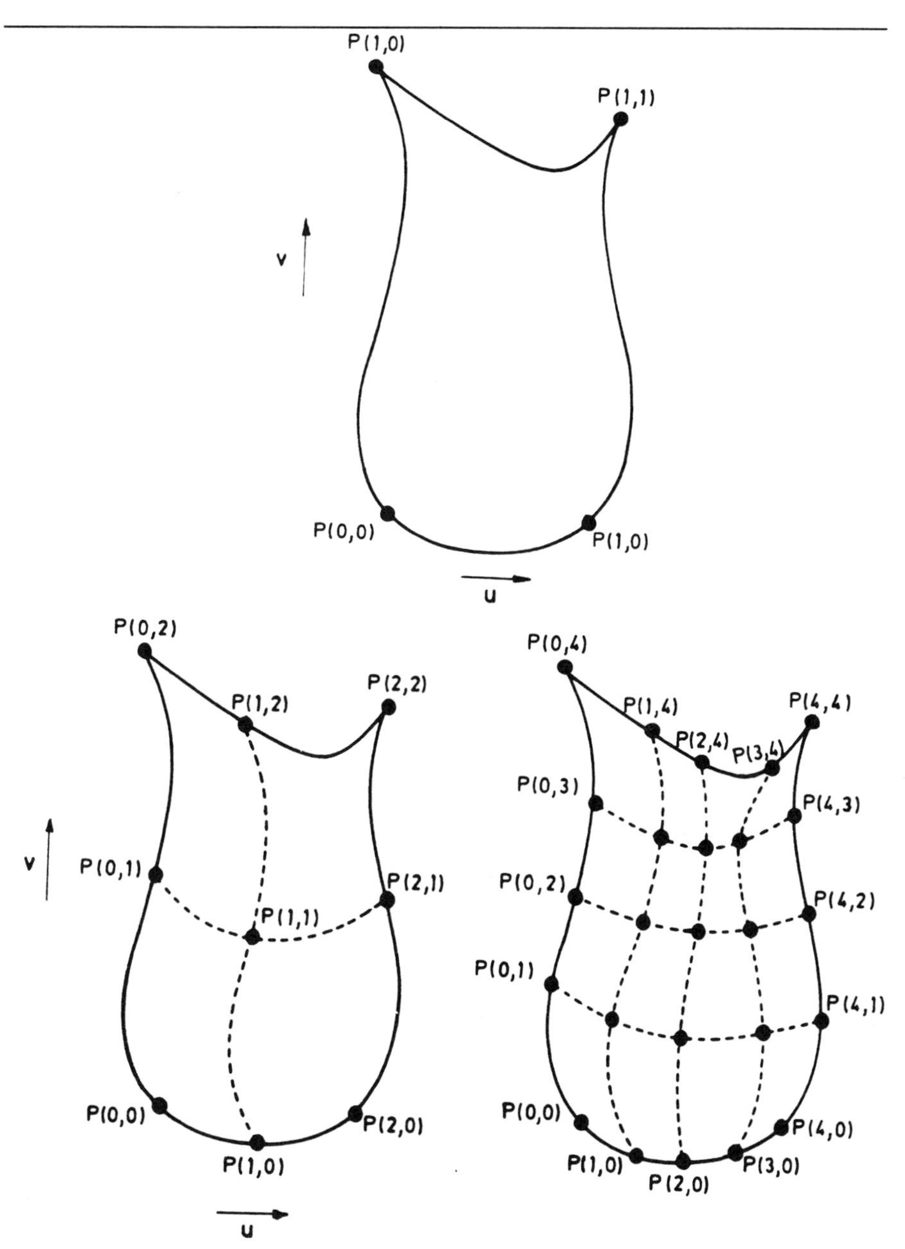

Figure 6.4. Level 0, level 1 and level 2 composite surfaces.

6.3.1. CALCULATING TANGENT VECTORS

Once the 4 knots that define the extremes of the rectangular defining grid are chosen, the tangent vectors in each of the parametric directions must be calculated. The contour of the region contains a series of 3-D data points obtained from stereo matching that define four boundary curves on the surface. These curves are approximated by a least square cubic polynomial parametrized by arc length which is then differentiated and scaled to yield tangent vector values for the knots. The scaling is necessary since the approximating curve and the defining parametric grid use different parameters.

6.3.2. CALCULATING TWIST VECTORS

The twist vectors are more difficult to estimate. In the non-parametric representation of a surface

$$z = G(x,y) \tag{6.3}$$

the cross derivative $\frac{\partial^2 z}{\partial x \partial y}$ measures the rate of change in the x direction of the slope of the surface in the y direction, or the twist in the surface. The parametric cross derivatives are related, but since the actual surface twist is found by ratios of the parametric derivatives, they can be an artifact of the particular choice of parameters. If the parametric directions on the surface are along the lines of curvature of the surface, then there is no twist in the surface and the twist vectors are zero.

In practice, if care is taken, these vectors can be set to zero with minor effects on the surface. This assumes that the parametrization of the surface has been chosen wisely, with corner knot points chosen at places of high curvature or discontinuity along the boundary and spaced uniformly in both parametric directions. Attempts have been made to estimate these vectors by sampling surface data at the corner points with the tactile sensor but the results were not useful. As the number of patches that interpolate the surface increases, the effect of these twists is reduced since they only need be computed at the four corners of the knot grid.

Selesnick [56] has suggested a method for computing the twists from surface data at the knots. His method relates the twist normal component to the Gaussian curvature of the surface which can be computed locally. Once the Gaussian curvature is computed, the component of the twist vector in the direction of the surface normal may be computed. This leaves the surface tangential components of the twist to be estimated, which can be done accurately with locally sampled data. For the purposes of this research, the twists are assumed to be zero.

6.4. BUILDING HIGHER LEVEL SURFACES

A level 0 patch is built from vision data only and is not an accurate description of the underlying surface. There are an infinite number of surfaces that can fit the sparse boundary contour that vision supplies. Further, the tangents which are estimated from stereo match points are inaccurate along contours that are horizontal due to the lack of stereo match points. What is needed is information in the interior of the region to supplement the boundary information. This information can be obtained by the active tactile exploration algorithms described in chapter 5. From a level 0 surface, a level 1 surface can be built that includes more surface information in the interior. Figure 6.4 describes the method of building higher level surfaces. A level 1 surface is formed by adding a tactile trace across the single surface patch defined in level 0, and a level 2 surface is formed by adding tactile traces to each of the 4 patches defined by level 1 creating a new surface with 16 patches. This method is hierarchical and general, allowing surfaces of arbitrary level to be computed. The only restriction is that the new composite surface is globally computed. This means that given a knot set at resolution $N \times N$, the new knot set will be at resolution $(2N-1) \times (2N-1)$, involving tactile traces in $(2N-2)\cdot(2N-2)$ patches. By using higher order polynomial surfaces, local adjustments in the patches are possible; however the extra computational burden is not warranted by using fifth degree or higher polynomials. In practice, a level 1 patch containing a 3 x 3 knot set and 4 patches shows good results.

Algorithm 6.3 describes the procedure for creating a level 1 surface from a level 0 surface. A level 0 surface has a 2 x 2 knot set and 1 patch, and a level 1 surface has a 3 x 3 knot set and 4 patches. The algorithm uses the Trace Surface algorithm (algorithm 5.2) to generate interior surface

---

Algorithm to create a new level 1 patch from a level 0 patch.

Input: Level 0 patch containing a 2 x 2 knot set.

Output: Level 1 patch containing a 3 x 3 expanded knot set.

1. Move the sensor to the point of surface contact determined by the Explore Region algorithm (algorithm 5.1).

2. Using the Trace Surface algorithm (algorithm 5.2) trace from the surface contact point to the midpoint of each of the boundary curves in the level 0 patch. The movement vector **M** in the Trace Surface algorithm is computed using the midpoint of each boundary curve as the goal point and the knot points at the end of each curve as the equal parameter spacing points.

3. Create a new knot set with the old knots as the corner knots of a 3 x 3 knot set. The initial surface contact point will become the knot in the center of the grid. The final contact point of each trace becomes the new knots in between the old 2 x 2 knot set.

4. Adjust the tangents at the corners of this new knot set by recomputing the cubic least square boundary curves between the old knots to reflect the added tactile information on the boundary curves. Differentiate the curves and scale the tangents to reflect the change in parametrization.

5. Add the tangents at the new knots by forming cubic least square polynomial curves from the tactile trace data. Differentiate the curves and scale the tangents to reflect the change in parametrization.

Algorithm 6.2. Create New Patch.

---

information. The traces begin at the point of surface contact found in the Explore Region algorithm (algorithm 5.1), which is the centroid of the region. The algorithm then traces in the direction of the midpoints of the level 0 boundary curves. The traces preserve the equal parametrization on the surface by using the knot points at the ends of the boundary curves to calculate the movement direction on the surface. The points reported during these traces are combined into cubic least square polynomial curves that are differentiated and scaled to calculate the tangential information needed at the boundaries. The tangents of the boundary curves computed from vision data are updated to include the new tactile information, which fills in areas that lack horizontal detail from the stereo process.

Figure 6.5 shows a level 1 patch of the body of a pitcher. This patch was built from real stereo contours and active tactile sensing of the interior of the patch. The vision modules supplied sparse 3-D contours for the region that were used to define an approach vector to the surface. The surface exploration algorithm moved the sensor onto the surface and using tactile feedback, traced the interior of the surface. These tactile traces were then integrated with the sparse stereo contours to create the level 1 patch. Further results from real data using active tactile sensing are reported in chapter 8. The method is able to accurately interpolate planar, cylindrical and curved surfaces.

It is important to note that the vision processes are supplying the justification for building smooth curvature-continuous surfaces from a region. If the region were not a smooth surface, then zero-crossings would have appeared inside the region, precluding the assumption of smoothness. The lack of zero-crossings, or the "no news is good news" criteria established by Grimson [21] supports this method and in fact is the reason it succeeds in interpolating the surfaces well.

6.5. BUILDING HOLE AND CAVITY DESCRIPTIONS

The Trace Hole/Cavity algorithm (algorithm 5.3) describes the method for tracing the contour of a hole or cavity with the tactile sensor. A hole or cavity is described by its approach axis and a planar cross section. The sensor reports points of contact as it moves on and off the surface surrounding the hole or cavity. This can be a noisy procedure as many of the tactile sensor's contacts become activated in a small tight area such as the hole in

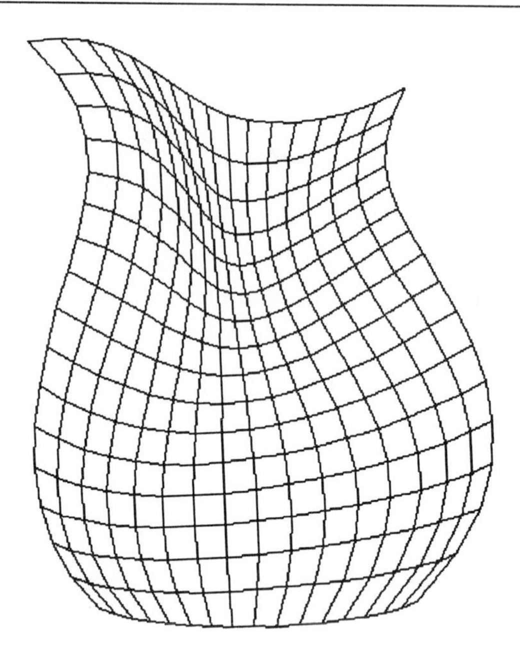

Figure 6.5. Level 1 surface of a pitcher.

the handle of a coffee mug. The poor spatial resolution of the sensor contacts also contributes to this phenomenon. The data is not continuous, but is a set of ordered contact points. Linking these points with line segments yields a curve that needs to be smoothed. The smoothing of each boundary curve is done by approximating the series of linked contour points with a smooth periodic spline curve. The periodic spline matches derivatives at the endpoints; this is important in creating smooth curves that are closed. Figure 6.6 shows a set of noisy linked sample points of a circular boundary curve and the boundary curve created by smoothing with splines. Once the curve is smoothed, a moment set is computed for the cross section bounded by the curve using the methods of chapter 2. The moments are important in determining the transformation between sensed and model coordinate systems.

6.6. SUMMARY

The integration of vision and touch is the cornerstone of the recognition process. This method allows full 3-D surfaces to be created from sparse vision and active tactile sensing. The method requires the use of the active tactile algorithms discussed in the previous chapter to control the movement of the arm and sensor as it traces surfaces and features on the object. The composite surfaces that are built from this method are smooth interpolants of the actual surface, able to be sensed at varying levels of resolution, and they are represented in an analytic form which allows simple computation of attributes for matching. The smoothness constraint is an outcome of the vision analysis which yields regions without interior zero-crossings to explore. Boundary curves of holes and cavities are also found through a combination of visual and tactile sensing. These curves are then smoothed to negate sensor noise effects and create an accurate boundary description.

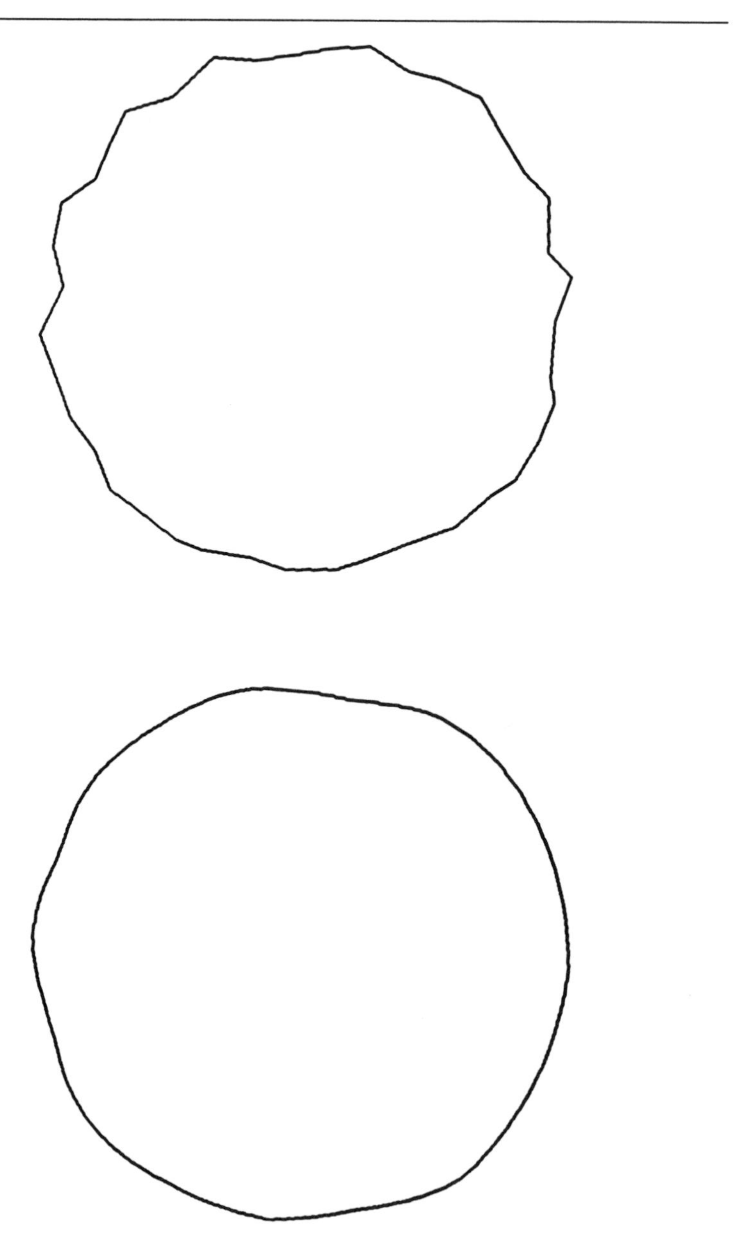

Figure 6.6. Sampled and smoothed boundary curves for a circle.

CHAPTER 7

MATCHING

7.1. INTRODUCTION

The low level vision and tactile algorithms provide a set of 3-D surface and feature primitives that are used by the matching routines to determine what the object is and to determine its orientation. The matching routines try to find an object in the model data base that is consistent with the surface and feature information discovered by the sensors. The intent is to invoke a uniquely consistent model from the 3-D surface and feature primitives discovered. If more than one consistent object is found in the data base, a probabilistic measure is used to order the interpretations. Once a consistent interpretation is found, a verification procedure is begun. This requires the matcher to calculate a transformation from the model coordinate system to the sensed world coordinate system. This transformation is then used to verify the model by reasoning about the slots in the model data base that are not filled. The initial choice of a model is made easier by the 3-D nature of the primitives, allowing matching of higher level attributes rather than sets of confusing and noise filled point data. The rules used for invoking a model are such that no a priori choice of features or surfaces is needed; all the structural parts of the model are candidates for matching. The object recognition system has no way of knowing what features or surfaces will be sensed from a particular viewpoint. It must be able to invoke a model based upon any identifiable part of the model [2].

The matching phase is the most difficult of all the modules since it requires the system to do high level reasoning about objects and their structure based upon incomplete information. The approach taken here is to develop a set of rules that will allow experimentation with different reasoning strategies to try to develop this capability in the system. The strategies and rules to be used are still under development and will require further research and are an obvious extension to this work. At present, a set of rules exist for the instantiation phase of matching. The verification phase is currently not implemented as an integral part of the system. Programs that carry out verification sensing have been developed and are demonstrated in chapter 8.

This chapter explains the strategies and techniques for matching developed so far and proposes directions for future research. Chapter 8 discusses the experimental results achieved with the rules and methods described below.

7.2. DESIGN OF THE MATCHER

An important design decision in building a matching system is when to invoke the higher level knowledge in the model. The information encoded in the model is rich and useful, and it would be helpful to the low level modules to have such information as early as possible. For example, if the first region explored by vision and touch is a hole, a possible strategy might be to search the data base of models and find all objects with holes that are consistent with the sensed hole. The reasoning modules could then try to find discriminating structures in the models with holes to suggest the next level of sensing. While it appears that humans may be capable of such reasoning, it clearly is beyond present machine capabilities to reason this strongly. The approach taken here is to sense as much as possible initially to try to limit the burden on the reasoning modules. If many primitives are found, the probability of a unique and consistent interpretation increases. The cost of this extra sensing is minimal. The discrimination that must take place to distinguish similar objects will no doubt cause most of the region sensing to occur anyway. Invoking the model later in the recognition process eliminates many blind alleys caused by reasoning with incomplete information at the cost of sensing up front. If a unique interpretation results from sensing all the regions, then the probability of a correct interpretation is increased. As Binford has stated in [7]:

> In machine perception, overwhelming verification of a correct hypothesis is typically inexpensive compared to the computation required to get to the correct hypothesis. These factors shift the utility balance toward getting data needed for a highly constrained decision. Very strong, relevant data are available if descriptive mechanisms can abstract them and interpretation mechanisms use them.

The implementation of the matcher consists of a set of Prolog goals that match sensed regions with model nodes. The model data base is implemented as a set of Prolog facts that are indexed in a hierarchical manner. The data base consists of eight kitchen objects: pitcher, mug, spoon, teapot, plate, bowl, drinking cup, pot. Four of these objects (pitcher, mug, plate, bowl) were used in experiments to test the matcher and its ability to correctly identify the objects. The other objects were included in the data base to see if the discrimination would work in certain selected cases discussed in chapter 8. The model data base is limited in size by the difficulty of the modeling effort and desire for efficiency in the matching routines. The matching routines described below are intended to show strategies and methods for model instantiation, computation of transformations from model to sensed world coordinates, and verification procedures. The discriminations produced by the matcher are meaningful because the discriminations are based upon actual sensed 3-D structure.

7.3. MODEL INSTANTIATION

The first phase of matching is to try to instantiate a model which is consistent with the sensory data. The rules for instantiation are based upon the sensed attributes of each region investigated by vision and touch. These regions may turn out to be surfaces, holes or cavities, and it is important that the instantiation rules not favor one or more of these access routes into the data base. The hierarchical nature of the model allows access to the model attributes at different levels depending upon the kind of sensory data produced.

The matching of sensed data against the model data base can be prohibitively expensive if all sensed regions must be matched against all model nodes. The instantiation phase tries to limit the number of feasible models quickly using easily computed criteria. Once the initial set of consistent interpretations is produced, more detailed matching occurs to try to determine a transformation from model to sensed coordinates. Finally, the verification will perform a new level of sensing to test the hypothesized model for consistency.

7.3.1. DISCRIMINATION BY SIZE

One of the benefits of using active tactile exploration is that physical size constraints can be used for global discrimination. Nevatia and Binford [43] and Brooks [13] have shown the utility of using physical size constraints in recognition tasks. The tactile sensor can be moved into the workspace to trace the global outline of the object to determine its bounding box. This is done by aligning the sensor vertically with the worktable and moving the sensor until it contacts the object. The sensor then moves around the object until it returns to its starting position. The granularity of the movement may be varied to obtain coarse measurements or produce finer detail. This is a simple, fast and effective procedure for limiting the initial search space of the object models. Any model whose bounding volume exceeds the sensed volume is rejected. This procedure also puts coarse bounds on the location of the object which can be used by the verification procedures later.

7.3.2. DISCRIMINATION BY GROSS SHAPE

Another simple discrimination test that is useful is discrimination by gross shape. The 3-D sensory data supplies information on features (holes and cavities) and surfaces. If the low level sensing discovers N holes, all models with N-1 holes can be rejected. This applies equally well to cavities. For surfaces, the criteria is more strict. Because the sensors discover patches of possibly larger surfaces, the surface type classifier is less robust. A curved surface in the model may have cylindrical regions, which may be sensed as a cylindrical partial patch. Therefore, gross shape discrimination must be conservative in matching curved surfaces. In the case of planar surfaces, discovery of a planar surface is a strong discriminant. The procedures used to classify surfaces are discussed in section 7.3.4.

7.3.3. FEATURE ATTRIBUTE MATCHING

Feature attributes are used as a discrimination tool to invoke a consistent model. Holes and cavities are modeled as right cylinders with constant planar cross section perpendicular to the cylinder's axis, occupying a negative volume. The constant cross section can be used to define a set of moments that can be used to match the cross section with a sensed feature. Moment matching was first described by Hu [30] who described a set of seven moment invariants involving moments of up to third order. These moments are simple to compute using the methods described in chapter 2. At the instantiation level two matching criteria are used. The first matches the moment M_{00} which measures the area of the planar cross section. For a match to be accepted, the sensed and model areas of the cross sections must be within a threshold. If this moment matches within the threshold, then the invariant $M_{02} + M_{20}$ is matched between sensed and model systems. This measure is scaled to reflect the difference in M_{00} when it is matched. In the case of cavities, an extra attribute of depth is available as a matching criteria.

Each feature is defined by its planar boundary curve and axis. The methods of moments was chosen for its simplicity of computation and matching. Other methods may be used besides the method of moments to match the curves. 2-D curve matching is a well studied problem. Other approaches are the curvature primal sketch of Brady and Asada [11] and the methods of curve matching developed by Kalvin et al [33] and Faugeras and Bhanu [6].

Once we have computed the moment set, the invariants can be used for matching. At the model instantiation phase, matches are only rejected if a strong rejection criteria exists since it is unlikely that a globally poor match will survive this culling. This discrimination becomes more and more robust as multiple features and surfaces are discovered.

7.3.4. SURFACE ATTRIBUTE MATCHING

The surfaces created from vision and touch need to be matched against the surfaces in the model data base. This problem is compounded by the fact that the surfaces created from vision and touch may be contained within or partially overlap the surfaces to be matched against in the model data base. There are two levels at which this matching takes place. In the instantiation phase, surfaces are matched according to global criteria described below. This phase tries to match surfaces by such attributes as area and type of surface. After a model has been instantiated, finer level matching is attempted to try to ascertain a transformation matrix between model and sensed object.

The initial phase of matching surfaces tries to match on two attributes, area and type of surface. The area criteria is useful in the context of posing initial consistent matches between sensed and model objects. The sensor is not capable of sensing accurately parts of the model with fine structure such as the handle of the mug. The area criteria effectively culls out small feature matching and leaves the task of larger shape correspondence. Any patch whose area is smaller than the sensed patch's area will be rejected as a match.

7.3.5. CLASSIFYING SURFACES

As described in chapter 2, the Gaussian curvature K is a measure which describes the local surface changes by means of a scalar. Using this measure, sensed surfaces can be classified as planar, cylindrical or curved. The procedure to do this iterates over the parametric surface at a specified sampling increment, computing the normal curvatures κ_{max} , κ_{min} in the principal directions and computing K as the product of these curvatures.

To classify a surface as planar, two criteria must be met. The Gaussian curvature computed over the surface must be within a threshold of zero everywhere and the surface must pass a planarity test. The test for planarity computes a least square plane $Ax + By + Cz + D = 0$ from a set of points on the surface. Residual distances for each point (x_i , y_i , z_i) in the set to the plane were computed from

$$R_i = (Ax_i+By_i+Cz_i+D)^2 \tag{7.1}$$

and a measure r of the planarity of the points was defined as:

$$r = \sqrt{\bar{R}} \tag{7.2}$$

where $\bar{R}$ is the mean residual in (7.1). If r is below a threshold, then the surface is classified as planar. Cylindrical surfaces are those with K=0 (within a threshold) and having a non zero κ_{max} or κ_{min}. Curved surfaces are computed similarly and have a nonzero value of K.

A method developed by Koparkar and Mudur [37] also can test for planarity directly from the surface patch equations. This result has related the planarity of a bicubic surface patch to the boundary curves of the patch. The method can be applied here to find out how planar a surface patch is. A Coons' patch can be defined as

$$\sum_{i=0}^{3} \sum_{j=0}^{3} A_i(u)\, B_j(v)\, Q_{ij} \tag{7.3}$$

where A_i and B_j are the blending functions of the patch and Q_{ij} are coefficients computed from patch data. If the blending functions are linear, and the patch corners are coplanar, then the patch is planar. By establishing small tolerances for coplanarity and linearity, the patch may easily be tested. This theorem states that the linearity of the patch is a function of the linearity of its blending functions, which are readily accessible and easily computed. Determining the linearity of the blending functions is accomplished by finding the curve maxima or minima measured from the chord joining the curve's endpoints.

Surfaces will match at this stage if two criteria are met. The first is that the sensed surface and matched model surface are of the same type as defined above. The second is that the sensed patch's area must be less than or equal to a model surface's area. If these criteria are met, then the surfaces are judged consistent in this first level of discrimination. The surface matches must then be relationally consistent as described below.

7.3.6. RELATIONAL CONSISTENCY

The set of possible consistent interpretations can be restricted further by maintaining relational consistency between the sensed regions and the model nodes. The relational constraint used is adjacency. If two sensed regions in space are physically adjacent, then the model nodes that these regions match with must also be adjacent. The list of potential matches generated from the surface and feature matching is further reduced by this method. For example, if a planar surface that is the lid of a teapot is sensed, it will match with both the bottom planar surface of the teapot and the lid. If the hole in the handle of the teapot is also sensed, an adjacency relation exists relating the two sensed regions (hole and lid). Sensing and matching the hole will cause the planar surface match with the bottom of the teapot to be rejected. It will be rejected because the model contains no adjacency relation between the hole and the bottom of the teapot. There does exist a model relation between the hole and the lid, and this is consistent with the sensed adjacency relation, causing this match to be accepted.

7.3.7. ORDERING MATCHES

The initial search for consistency is done by creating lists of all consistent matches between a set of sensed regions and the nodes in the data base. The sensed regions are described by a data structure that contains a list of cavities, a list of holes and a list of surfaces that have been discovered. These lists are then compared with each object node, trying to match lower level surface and feature nodes with the sensed data. The output of this matching is sets of consistent matches ordered by a combined probability-complexity measure. If the particular view that is presented to the sensors is rich in structure that can be sensed, then the matching described above is strong enough to invoke a unique consistent interpretation. If the view does not provide strong discriminating features and surfaces, then the consistent matches must be ordered for later verification.

There are two cases to consider in ordering matches. The first is that of matches that are consistent within the same model object. For example, consider an object with two cylindrical surfaces, equal in area. A sensed cylindrical patch may match either of these model nodes. Both matches will be accepted, but a preference will be given to the match of the model node containing the higher probability. The probability measure is an aid for recognition in which the components are ordered as to their likelihood of being sensed. High priorities are assigned large components or isolated components in space that protrude (handles, spouts). Obscured components, such as a support surface of an object are assigned lower priorities. The probabilities do not preclude recognition but simply give a preference for one set of potential matches over another. The probability measure is normalized across all objects so that each object's surface and feature probabilities sum to 1.

The second case is that of consistency across different model objects. Given a set of consistent object matches a strategy for determining which object is present is needed. The set of consistent interpretations needs to be partitioned in some manner. In general, determining these partitions dynamically is very difficult. A possible solution is to partition the objects *a priori;* however, the space of possible consistent interpretations is too large for this to be an effective strategy. The strategy used here is to search for object complexity. To implement this strategy, a complexity attribute is attached to each object model which is the number of components and features in the model of the object. The normalized probability measure computed from matches within each object is multiplied by the model complexity and the matches are ordered by this measure. Given two matches of equal probability, the more complex object will be preferred, and verified first. This choice was made for two reasons. First, the sensors are more capable of finding the presence of a surface or feature than the absence of one. Secondly, finding a surface or feature not only helps the discrimination but quantifies it through sensing the surface or feature attributes.

7.4. VERIFICATION

Verification can be viewed as slot filling, where the instantiated model's nodes are either filled, representing a sensed match, or unfilled. Verification becomes a process of reasoning about unfilled slots. The first step in this process is to compute a transformation between the model coordinates and sensed world coordinates. Once this transformation is computed, verification sensing can be carried out, using the sensors to discover unsensed or occluded structure.

7.4.1. COMPUTING MODEL-TO-SCENE TRANSFORMATIONS

Once a model is instantiated, a transformation between model coordinates and sensed world coordinates must be computed. This transformation will allow the knowledge embedded in the model coordinate frame to be used in the sensed world frame. By transforming model surfaces and features to the sensed world frames, verification of unrecognized slots in the model can proceed since their assumed location is now computable with this transformation. This knowledge enables the sensors to explore regions that were not seen in the initial sensing and to explore visually occluded areas with tactile sensing. The transformation may be computed with feature information or surface information. In some cases, a partial transformation may be computed that will allow further sensing.

7.4.2. MATCHING FEATURE FRAMES

Each feature in the data base is associated with a coordinate frame. This allows the feature to be defined in object-centered terms rather than arbitrary model coordinates. Once the models and their frames are developed, mappings from one feature frame to another are readily computable. Figure 7.1 shows the frames C_m and H_m which are object-centered frames defined for a coffee mug's cavity and hole in the model coordinate system. The relative transform between the hole frame and the cavity frame R_{hcm} can be defined as:

$$C_m = H_m : R_{hcm} \tag{7.4}$$

$$R_{hcm} = H_m^{-1} : C_m \tag{7.5}$$

Similarly, the transformation from modeled cavity to modeled hole R_{chm} is:

$$R_{chm} = C_m^{-1} : H_m \tag{7.6}$$

Because these are relative frames, discovering one of the model frames in the sensed coordinate space will define the other feature in the sensed coordinate space. Assuming we know the match between the hole in sensed world coordinates with frame H_s and the model hole with frame H_m then the cavity in sensed world coordinates is defined by frame C_s:

$$C_s = H_s : R_{hcm} \tag{7.7}$$

The determination of the new feature frame in sensed world coordinates is important to the verification process. If an unfilled feature slot is seen, then the feature's frame in sensed coordinates is available through the relative frame mapping. The frame for a feature defines the axis of the hole or cavity in sensed world coordinates which is then used as an approach vector to sense the unseen feature even if it is occluded.

In some cases, feature frames are only partially defined. This is the case with rotationally symmetric features such as a circular cavity or hole. The approach axis of these features is well defined, but the principal axes of inertia are not. However, the frame matching technique discussed above can still determine within this rotational parameter the new sensed frame. If occlusion information is also included, the new frame can be constrained to lie within a certain rotational range around the approach axis, allowing occlusion sensing to take place in order to find features that cannot be seen. An example of this is given in the next chapter, where the tactile sensor is able to sense a visually occluded hole.

7.4.3. MATCHING SURFACE FRAMES

Matching of surfaces is more difficult because a unique surface frame is not as easily sensed as a feature frame. Planar and cylindrical surfaces have one well-defined frame vector which is the plane's normal and the cylinder's axis. Curved surfaces in general do not have any such natural, embedded frame vector. In the case of planar and cylindrical surfaces, the one axis which is defined will allow defining the transformation up to a rotational parameter about that axis and a translation. In the case of the plane, the plane's centroid is also computable and this will supply the translational component of the transformation. This can be used in conjunction with other feature and surface matches to constrain the sensed frame.

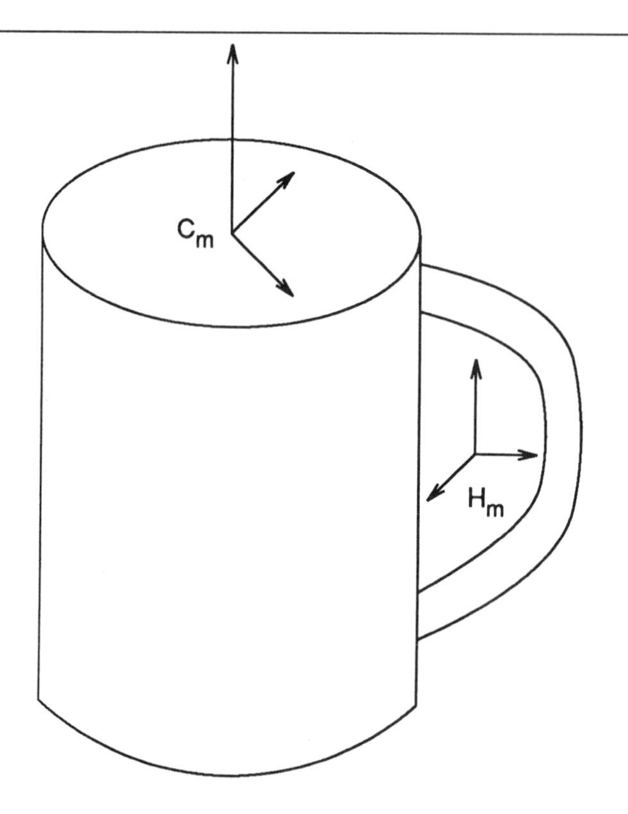

Figure 7.1. Coordinate frames for the features of a mug.

Curved surfaces have no embedded frame information that is unique that we may exploit for arbitrary surface frame matching. Unlike a planar or cylindrical surface, an arbitrary curved surface has fewer invariants such as a normal to a planar surface or an axis of a cylinder that can be matched against. One approach, implemented by Potmesil [51], is to generate point matches on the surface and try to iteratively compute the transformation matrix. Potmesil matched bicubic patch descriptions in order to build 3-D models of objects from different viewpoints. His method was to choose an initial set of point matches (four are needed) and compute the transformation from one patch to another and then test for correspondence. The method worked reasonably well but was slow and used an arbitrary evaluation function. No attempt was made to implement this method here due to the excessive time of execution which precludes its use in a robotics environment. Compounding this problem is the surface subset problem: the sensed surface is in general a subset of the larger model surface. Therefore the point matches must be chosen from a potentially larger set of points.

The analytic nature of the surfaces created from vision and touch allows computation of differential geometry measures such as lines of curvature, principal directions, and Gaussian curvature. Brady, Ponce, Asada and Yuille [12] have suggested that certain lines of curvature that are planar might be significant in terms of recognizing structure. For example, the only planar lines of curvature on an ellipsoid are the lines formed by the intersection of the symmetry planes with the surface. Discovery of lines such as these is feasible with the representation used, and may lead to more robust recognition methods for curved surfaces.

7.4.4. VERIFICATION SENSING

Once the transformation relating the modeled to sensed coordinates is computed, feature and surface locations in the model can be related to the sensed world coordinates. The location and approach axes of holes and cavities can be computed from these transformations and used to guide the tactile sensor to verify the feature's existence. In particular, occluded features may be sensed and verified in this manner. Because their approach axes and centroids are well defined by the transformation, blind tactile search can succeed.

Tactile sensing of visually occluded surfaces is difficult. The integration of vision and touch to build surfaces described in chapter 6 works precisely because both modalities are being used. The vision guides the tactile tracing, establishing starting and ending conditions on the trace. Attempts to use touch alone to build surfaces that are visually occluded will not work without the extra information supplied by vision. Blind touch can only sense discontinuities and presence or absence of surfaces. Further, because the touch is blind, the relation of these surfaces to the object structure is unclear. It is not possible to build a patch description as is done in the visible parts of the scene.

Verification can be time consuming if all model slots are to be filled. The hierarchic nature of the models supports different levels of verification sensing. If a component slot is filled because a surface of that component was matched, we can decide to accept the component as verified or do further sensing on any other surfaces that make up this component. If the model instantiated is unique, then lower levels of sensing may not be necessary. If the instantiation is not unique, then going to the next level of the hierarchy of slots to perform more sensing may be called for. Confidence levels for verification can be set up in this manner, suggesting different levels of acceptance and further sensing to be carried out.

7.5. SUMMARY

Matching is the last step of the recognition process. It has two components, instantiation and verification. Instantiation tries to find consistent interpretations from the sensed data using rules. Once a model is instantiated, verification computes a transformation from model to scene, allowing further sensing to take place and support or reject a hypothesis. Design decisions need to be made as to the levels of sensing and matching criteria that need to be established for each phase of the process. Matching is done at a coarse level to try to quickly reduce the number of feasible models. Matching is based upon attributes of surfaces and features discovered by the integration of vision and touch. If the imaged scene is rich with structure that can be sensed, instantiation of a unique model is likely. If a unique instantiation is not possible, then the possible objects are ordered by probabilistic and complexity measures.

The transformation from model to sensed coordinates may be computed from either feature or surface information. Features have an embedded coordinate frame that simplifies the computation of this transformation. Surfaces may only supply partial information about the transformation. Partial transformations will still allow further verification sensing to take place.

CHAPTER 8

EXPERIMENTAL RESULTS

8.1. INTRODUCTION

This chapter details the experiments that were conducted to test out the approaches developed in the previous chapters. The main intent of these experiments is to show that tactile sensing can reliably complement machine vision sensing and that together these two sensing modalities can create robust and accurate 3-D descriptions of surfaces and features. In particular, experiments have been run on scenes that would severely test the recognition capabilities of a machine vision system. The experiments begin with recognition of simple objects in regular poses and then move on to increasingly more complex objects in different poses. Finally, the results of an experiment to use verification sensing of a visually occluded feature is discussed.

In each of the experiments, a single object is imaged by the stereo cameras. The 2-D vision algorithms identify regions of interest and the 3-D vision algorithms generate a sparse set of 3-D contours of the regions. These sparse contours are able to orient and guide the tactile sensor mounted on the manipulator arm in an exploration of each region, classifying it as a surface, cavity or hole. The integration algorithms are then used to build level 1 bicubic surface patches of each surface region and smoothed boundary traces of each hole or cavity feature. The quantified level 1 surfaces and features are then used to find an object model in the data base that is consistent with the sensed data. Once this occurs, an estimate of the object's orientation in space is computed.

The experiments reported are all run using real objects and real data from noisy sensors. The nature of robotic sensing is that sensors produce errors; any methods developed for tasks such as these must be robust enough to succeed in such an environment. In particular, the tactile sensor is a relatively crude device in terms of spatial resolution. However, the integration of both sensing modalities is extremely helpful in making the system more robust, since neither modality is used exclusively. The experimental results show that the discovery of 3-D structure is extremely important in performing a recognition task, making the matching phase of the recognition cycle more accurate. Each of the experiments will be described in detail below.

8.2. EXPERIMENT 1

The first experiment was to recognize a salad plate which is a regular planar object. The digital images and zero-crossings are shown in figure 8.1 and the stereo matches in figure 8.2. The images yielded few feature points that could be matched to determine depth as expected with a smooth homogeneous surface. The stereo matcher was only accurate in matching zero-crossings up to 65° from vertical, yielding sparse and incomplete depth information. An image such as this would pose large problems for a vision system alone; the data is too sparse to support a consistent visual hypothesis. The region analysis revealed only a single region to be explored which was the central area of the plate. The tactile system explored the plate and built the surface description shown in figure 8.3 by integrating the touch and vision data into a level one surface description. The surface was sampled at small intervals in parameter space calculating the maximum and minimum Gaussian curvatures (K_{max} , K_{min}), the area of the surface and the root mean residual r which tests for planarity. Table 8.1 shows this analysis along with the analysis of the modeled salad plate's surface. [2]

The matching algorithm was able to match the surface with the model of the salad plate in the data base. An analysis of the surfaces in the model data base established a threshold value of ± 0.001 for determining zero Gaussian curvature. The surface patch's Gaussian curvature was within this threshold implying a planar or cylindrical surface, and the residual measure r was small confirming a planar surface. All of the objects in the data base except the cereal bowl have at least one planar surface, and are potential matches; however, when the areas were compared, the matcher was able to discriminate among these objects and choose the plate. Figure 8.4 shows the computed surface normals on the level 1 surface, verifying its planar appearance.

The normal of the least square plane fitted to the surface was the estimate for the orientation of the object. No other orientation parameters were available since the plate was symmetric about its planar axis. The difference in orientation between the level 1 surface's normal and the plate's normal, measured by taking the dot product of the normals, was approximately 6°.

Determining if a surface is planar is a strong constraint. It allows determination of an orientation in space and establishes a firm matching criteria. The use of touch in this experiment verified the planar nature of a surface. The sparse visual cues were not dense enough to support this conclusion from vision alone.

---

[2] All surfaces are shown orthographically projected, with no hidden lines removed. The lines on the surfaces are lines of constant parameterization, computed by holding one parameter of the surface constant and iterating over the other parameter.

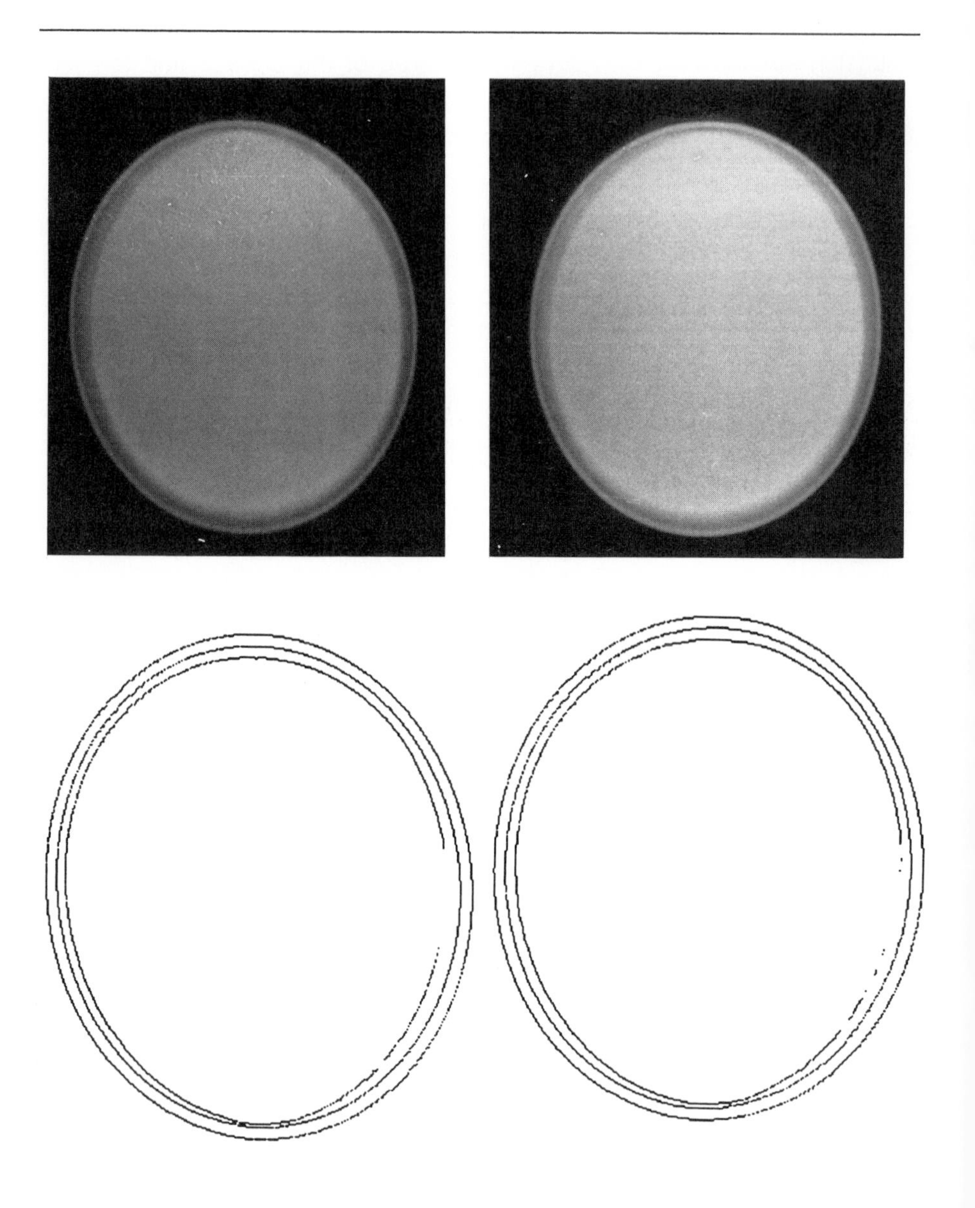

Figure 8.1. Digital images and zero-crossings for the plate.

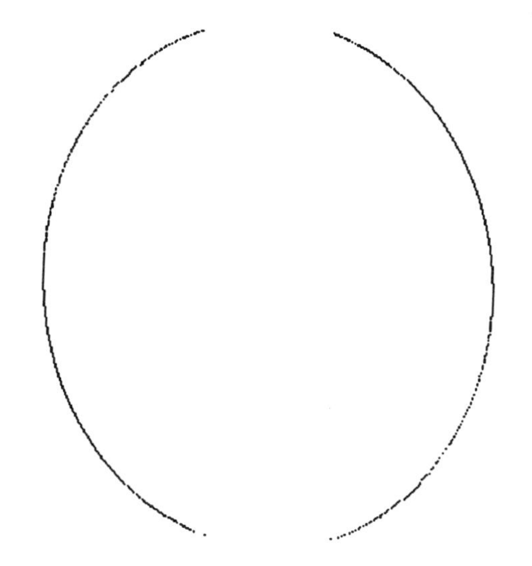

Figure 8.2. Stereo match points for the plate.

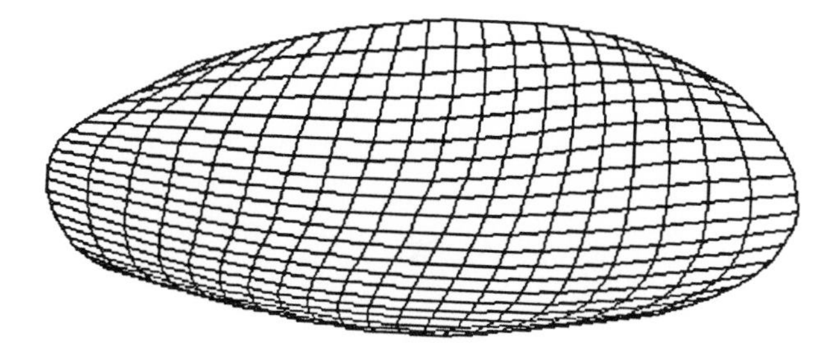

Figure 8.3. Level 1 surface for the plate.

Surface Analysis, plate				
Surface	Area	K_{max}	K_{min}	r
sensed	29470	.000896	-.000053	4.71
model	28496	.000326	-.000209	4.81264

Table 8.1. Surface analysis of the sensed and model plates.

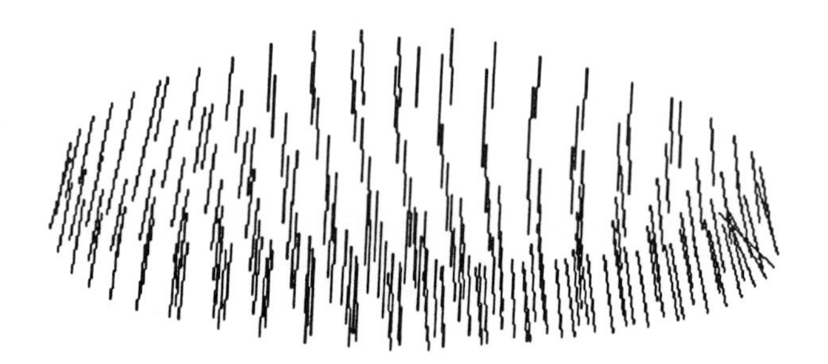

Figure 8.4. Surface normals on the plate.

8.3. EXPERIMENT 2

The second object imaged was a cereal bowl. The digital images and zero-crossings are shown in figure 8.5 and stereo matches in figure 8.6. The images are very similar to the plate in experiment 1. The only depth cues are monocular, where small shading gradients exist but which elude the zero-crossing edge detector. If surface reflectance and lighting were known, a possible method of shape reconstruction would be shape from shading. However, these constraints are unknown in our case. This is an excellent example of the discriminatory power when tactile sensing is added to vision. The region analysis yields one region to explore with the tactile sensor. Upon exploration, a level 1 surface of the bowl was computed and is shown in figure 8.7. Figure 8.8 is a cross section through the level 1 surface showing the computed surface normals. The normals reflect the hemispherical nature of the bowl.

The tactile sensor did not find a surface until it had passed 40 mm beyond the plane of the region's contour determined from vision. This prompted a cavity trace in addition to the surface trace. The cavity had a sensed depth of 40m m which was consistent with the model depth of 45 mm. Table 8.2 shows the moment sets for the sensed and model cavities which were also compatible.

The matcher tried to match the surface and the cavity with an object in the database. The surface is neither planar nor cylindrical since its Gaussian curvature is above the established zero threshold (table 8.3). However, the matcher chose the bowl because the level 1 surface's area was compatible with the bowl and the sensed cavity made the match unique. The cavity sensing yielded an estimate of the cavity's axis vector that was determined by the normal to the cavity cross section. The angular difference between the actual cavity axis and the sensed axis was approximately 5°.

The initial visual data for experiments 1 and 2 were almost identical. Only through the use of touch sensing did the surface's depth become apparent. The discovery of a cavity allowed the system to discriminate between two potential surface matches. The combination of surface and feature information reduces the likelihood of multiple consistent models being found.

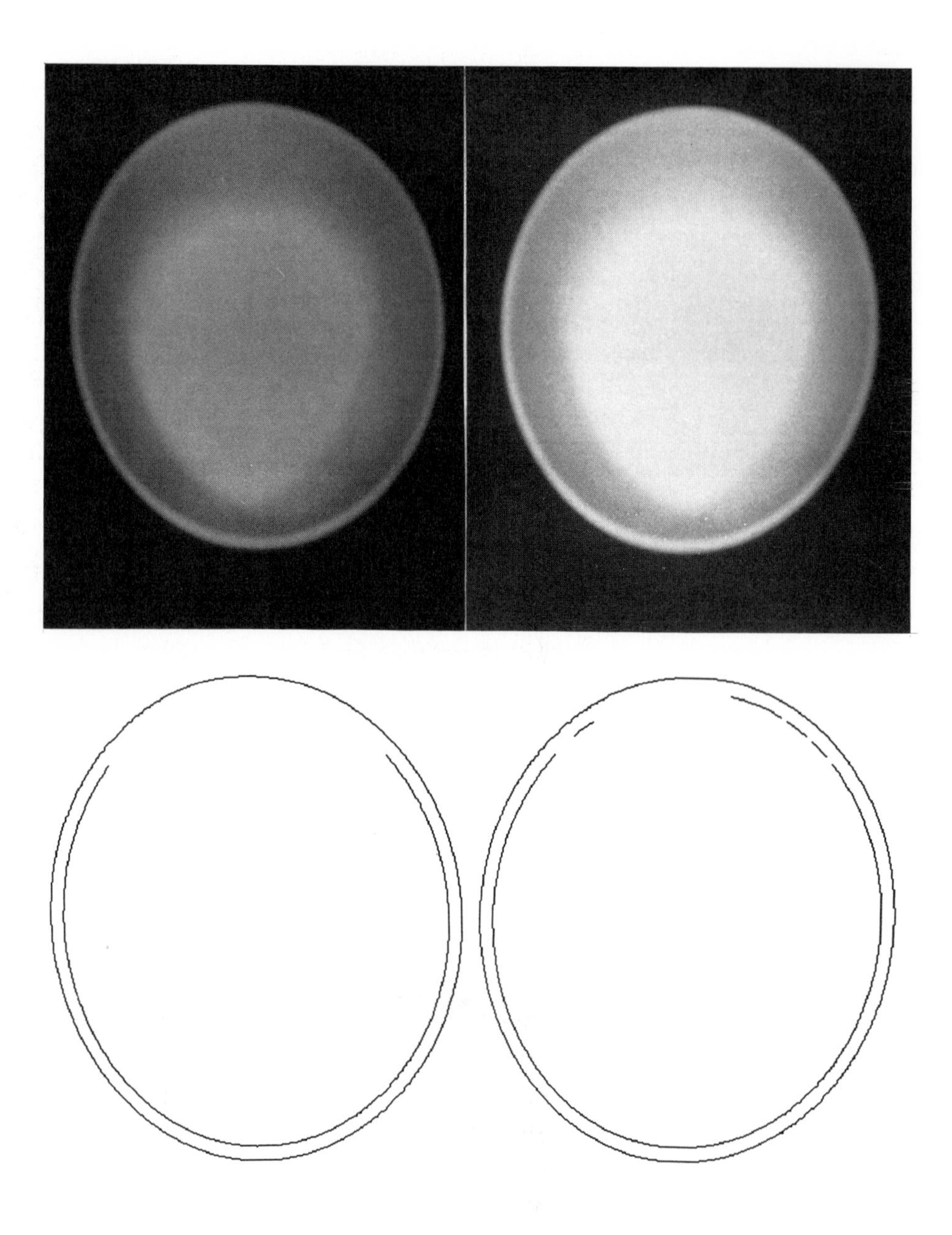

Figure 8.5. Digital images and zero-crossings for the cereal bowl.

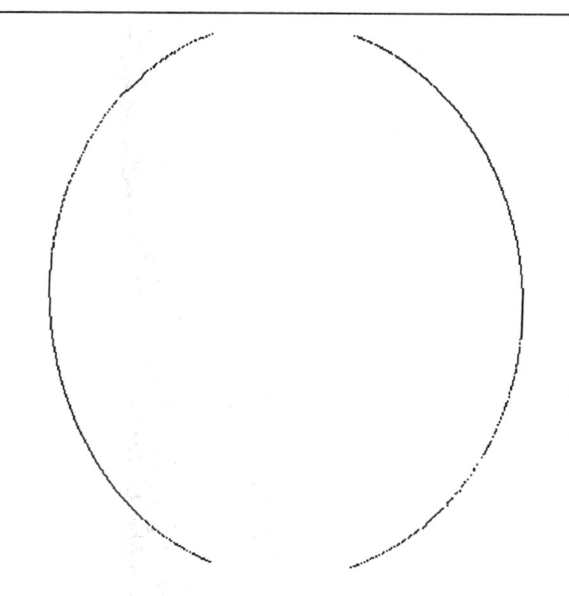

Figure 8.6. Stereo matches for the cereal bowl.

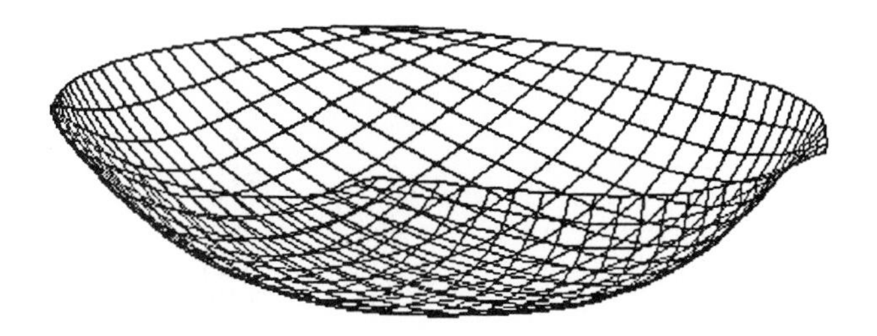

Figure 8.7. Level 1 surface for the cereal bowl.

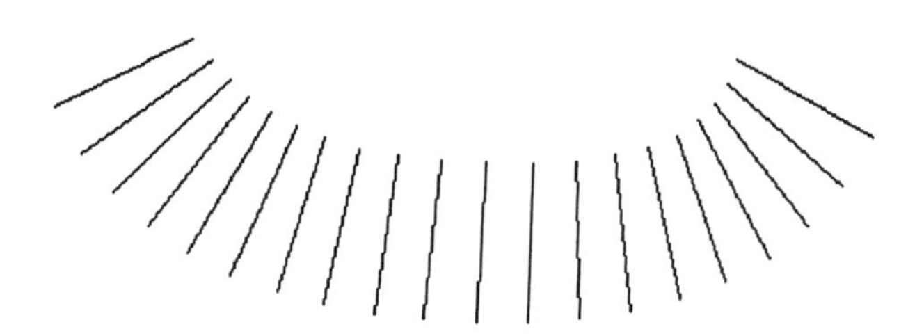

Figure 8.8. Cross section of level 1 bowl surface with normals.

Moments, cereal bowl cavity				
cavity	M_{00}	M_{20}	M_{02}	$M_{20}+M_{02}$
sensed	17068	21927264	24656910	46584174
model	19378	29882396	29882396	59764792
scaled sensed	19378	28263552	31781979	60045532

Table 8.2. Moments for sensed and model bowl cavity.

Surface Analysis, cereal bowl				
Surface	Area	K_{max}	K_{min}	r
sensed	41218	0.000949	-0.003550	14.9788
model	60616	0.001290	0.000432	17.2152

Table 8.3. Surface analysis of the sensed and model bowls.

8.4. EXPERIMENT 3

The third experiment imaged a coffee mug in which the hole, cavity, handle and body of the mug were all visible. The digital images and zero-crossings are shown in figure 8.9 and the stereo matches in figure 8.10. The region analysis yielded 4 separate regions to explore. The first region explored was the cavity. The sparse 3-D contours obtained from the vision algorithms determined an approach axis for the tactile sensor to the region. Upon tactile exploration, contact with a surface occurred only after the sensor had passed a significant distance past the contour points plane, implying a cavity. The cavity was then quantified by tracing the boundary and creating a smoothed boundary curve. Figure 8.11 shows the tactile sensor tracing the cavity of the mug, figure 8.12 shows the smoothed boundary curve computed from the tactile trace of the cavity and table 8.4 shows the computed moment set for the planar cross section of the sensed cavity. The second region explored was the mug's main body for which a surface patch was built as shown in figure 8.13 . This surface patch is a level 1 patch built from vision and touch and very closely approximates the cylindrical surface of the mug. The geometric analysis of the patch is shown in table 8.5 . The analysis of the patch shows its Gaussian curvature to be within the specified threshold of zero Gaussian curvature. The patch is not planar since its residual value from the fitted plane is too large. This leaves the choice of the surface to be a cylinder. The cylinder's axis can be determined by finding the lines of minimum curvature on the surface. The lines of minimum curvature determine one of the principal directions on a cylindrical surface. They are uniquely defined except in the case of *umbilic points* which are points on the surface where the curvature in each of the principal directions are equal (spheres and planes are entirely composed of umbilics). Figure 8.14 shows the surface normals computed for the level 1 surface, revealing the cylindrical nature of the sensed surface.

The hole was found after the Explore Region algorithm penetrated the region defined from vision processing with the tactile sensor and did not contact a surface (figure 8.15). The Trace Hole/Cavity algorithm traced the hole, and the smoothed boundary curve shown in figure 8.16 was computed from the contact points on the hole's boundary. Table 8.6 shows the computed moment set for the traced hole.

The matcher was presented with an abundance of sensed region information to try to instantiate a model. The cylindrical surface that was computed matched a number of objects in the data base (pot, coffee mug, drinking glass) as did the cavity (drinking glass, coffee mug). However, the sensed hole provided a final bit of discrimination evidence. The hole was not found in the drinking glass (an identical object in the data base to the mug but without a hole or a handle) but matched with the coffee mug, yielding a unique choice of object. This particular view was rich with information. Because we are able to sense actual 3-D structure, robust discrimination is possible.

The determination of multiple features and surfaces provided a method of verifying the system's estimate of the coffee mug's orientation in space. The cylindrical surface axis and the cavity axis are parallel in the model and the sensed and model axes differed by less than 5°. The agreement is quite close, showing the ability to determine orientation from both surfaces and features that is consistent with the model, creating confidence in the initial hypothesis.

One problem involved the tactile sensor's resolution. The handle of the mug is too small and fine for the sensor to adequately build a patch description. It could be verified as a surface with the sensor, but attempts at building a patch description failed due to the sensor's much larger size. However, even though the surface could not be quantified, the determination that the region is a surface is still useful in establishing matching criteria.

This experiment shows the many ways an object can be recognized. Holes, cavities and surfaces are all able to be used to both recognize and correctly identify orientation parameters for the objects. This is important in that certain viewing angles may present a confusing region that cannot be sensed accurately. However, if one of the regions is able to be sensed accurately, then a partial match can be established leading to later recognition.

Figure 8.9. Digital images and zero-crossings for the coffee mug.

Figure 8.10. Stereo matches for the coffee mug.

Figure 8.11. Tactile sensor tracing the coffee mug cavity.

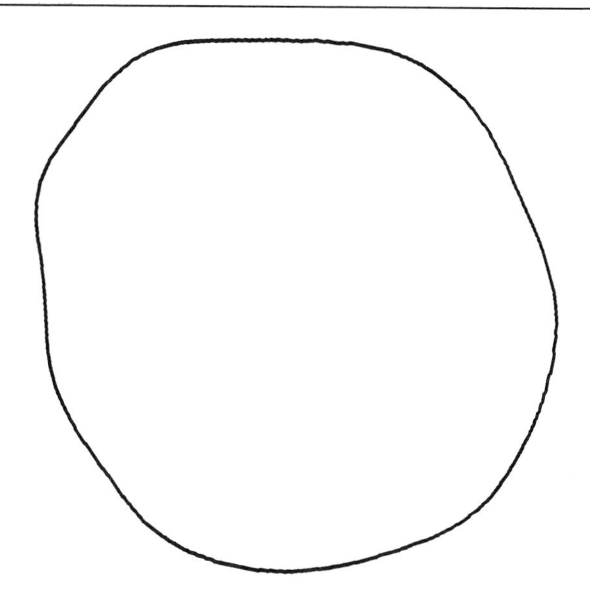

Figure 8.12. Smoothed boundary curve for coffee mug cavity.

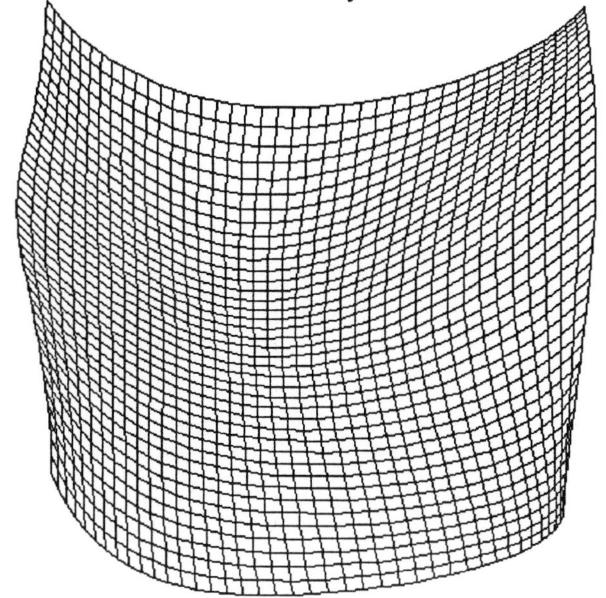

Figure 8.13. Level 1 surface for the coffee mug body.

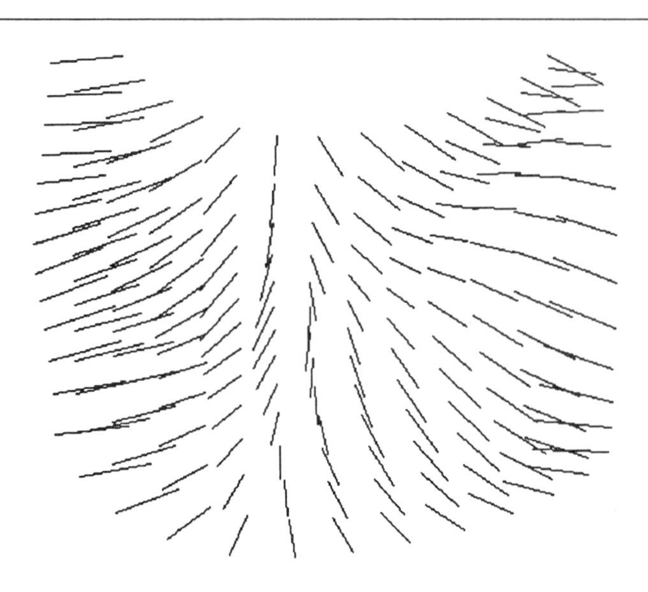

Figure 8.14. Surface normals for the coffee mug body.

Figure 8.15. Exploring and tracing the coffee mug hole.

Moments, coffee mug cavity				
cavity	M_{00}	M_{20}	M_{02}	$M_{20}+M_{02}$
sensed	4383	1485060	1583911	3068972
model	4758	1802083	1802083	3604167
sensed scaled	4758	1750112	18666606	3616718

Table 8.4. Moments for sensed and model coffee mug cavity.

Surface Analysis, coffee mug body				
Surface	Area	K_{max}	K_{min}	r
sensed	9598	0.000492	-0.00062	11.18
model	22078	0.00	0.00	9.73

Table 8.5. Surface analysis, sensed and model coffee mug bodies.

Moments, coffee mug hole				
hole	M_{00}	M_{20}	M_{02}	$M_{20}+M_{02}$
sensed	1011	152109	44729	196839
model	1296	187673	148729	336402
sensed scaled	1296	249843	73470	323313

Table 8.6. Moments for sensed and model coffee mug hole.

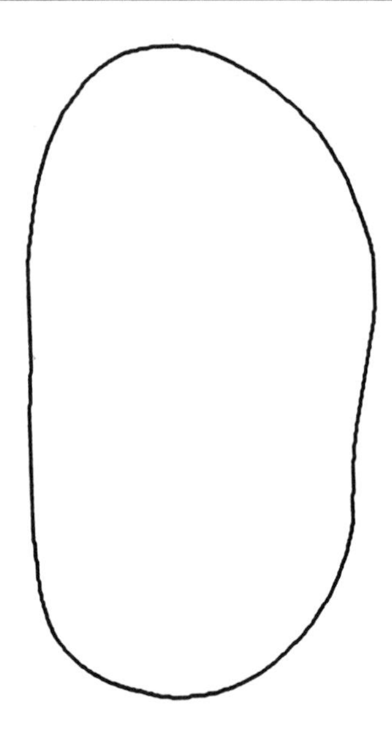

Figure 8.16. Smoothed boundary curve for coffee mug hole.

8.5. EXPERIMENT 4

The purpose of experiment 4 was to see if the system could determine if the mug was cavity-side up or bottom-side up. Visually these two positions are very similar. Only by exploring the region with the tactile sensor can the surface or cavity be distinguished. To perform this experiment, the mug was actually in the same upright position as in experiment 3. The visual analysis was the same as for experiment 3, but a thin plate conforming to the bottom surface of the mug was placed over the cavity opening during the tactile sensing. The tactile sensor reported a surface rather than a cavity. The surface is shown in figure 8.17 and the surface analysis is in table 8.7, revealing a surface with zero Gaussian curvature and a small value of r, confirming its planar shape. Figure 8.18 shows the surface normals verifying the planar analysis. The sensed plane's normal vector and the actual plane's normal vector agreed within 3°.

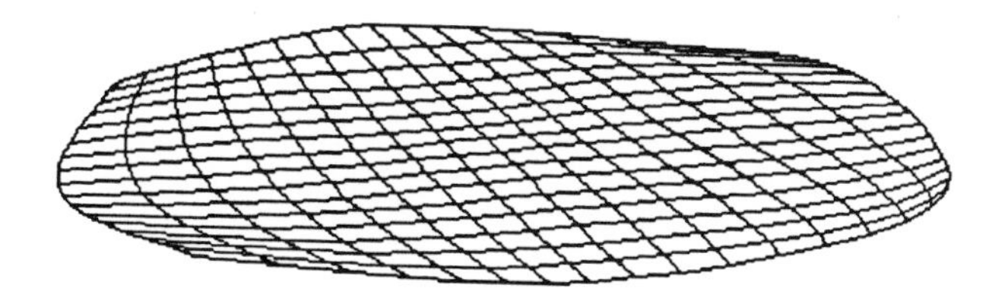

Figure 8.17. Level 1 surface for coffee mug bottom.

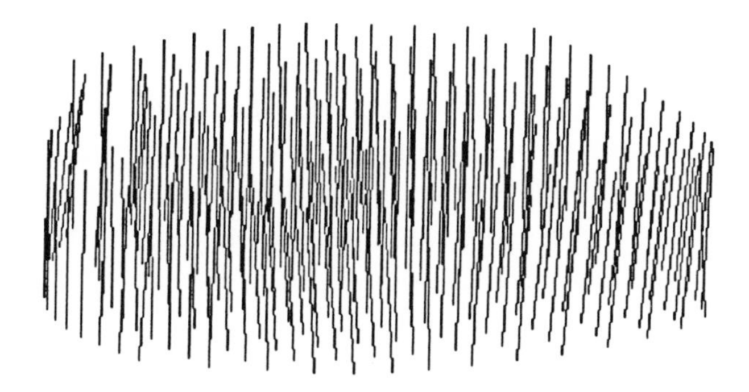

Figure 8.18. Surface normals for coffee mug bottom.

Surface Analysis, mug bottom				
Surface	Area	K_{max}	K_{min}	r
sensed	4790	0.000534	0.000299	0.577
model	5024	0.00	0.00	0.00

Table 8.7. Surface analysis, sensed and model mug bottom.

8.6. EXPERIMENT 5

In this experiment, a pitcher was imaged from the side. The digital images and zero-crossings are in figure 8.19 and the stereo matches in figure 8.20. The first region explored was the cavity, found by moving the probe along the normal to the cavity's contour plane and finding no contact within a specified distance of movement beyond the contour plane. The contour reported from this sensing was not accurate due to the shape of the cavity and the resolution of the sensor. The pitcher cavity has a variable cross section along its axis; it is not constant as the model expects. This prevents the Trace Hole/Cavity algorithm from properly tracing the cavity. The trace underestimated the cavity area by being below the plane of the wide mouth opening by a small amount of distance which yielded a different boundary curve. A solution to this problem is to modify the Trace Hole/Cavity algorithm to follow edge discontinuities; however the sensor being used has difficulty following fine changes in surface structure such as this due to its poor spatial resolution. A more accurate sensor would allow cavities such as this to be traced, extending the range of the objects that can be modeled and recognized.

The next region sensed was the main body surface of the pitcher. This surface is doubly curved and has twisted space curves for boundaries. However, the vision and touch routines were able to build a quite accurate level 1 surface which is shown in figure 8.21. This surface was built from sparse 3-D contours and a level 1 tactile trace across the region. The doubly curved nature of this surface can cause problems for systems that try to recreate this surface with either polygons or quadric surfaces; the composite Coons' patch technique is able to recreate the surface accurately and from a minimum of sensory data.

The Gaussian curvature (table 8.8) ranges from positive to negative on this surface describing a surface with hyperbolic and elliptic points. Figure 8.22 shows the surface normals computed from the surface and figure 8.23 shows the principal directions on the surface. The Gaussian curvature analysis rules out all planar and cylindrical patch matches. Thus the pitcher body is matched in the model when the further constraint of surface area is considered.

The hole is an excellent discriminating feature between the pitcher and the mug. The tactile routines were able to sense the hole and compute its moment set (table 8.9). A problem due to the relatively large tactile sensor size is its inability to accurately sense the point of the handle. The sensor is physically too large to fit into this space, and it is unable to sense the point of the handle. The area of the cross section is underestimated because of this but is still within a threshold of the model area. The smoothed boundary contour of the hole is shown in figure 8.24.

The system was able to discriminate in this experiment based upon surface differences and feature differences. The recognition was able to be done even though the cavity trace was unsuccessful. This is due to the fact that 3-D structure was being sensed and partial matches of this structure were strong. The discovery of the curved surface and the hole allowed the matcher to uniquely instantiate the pitcher model.

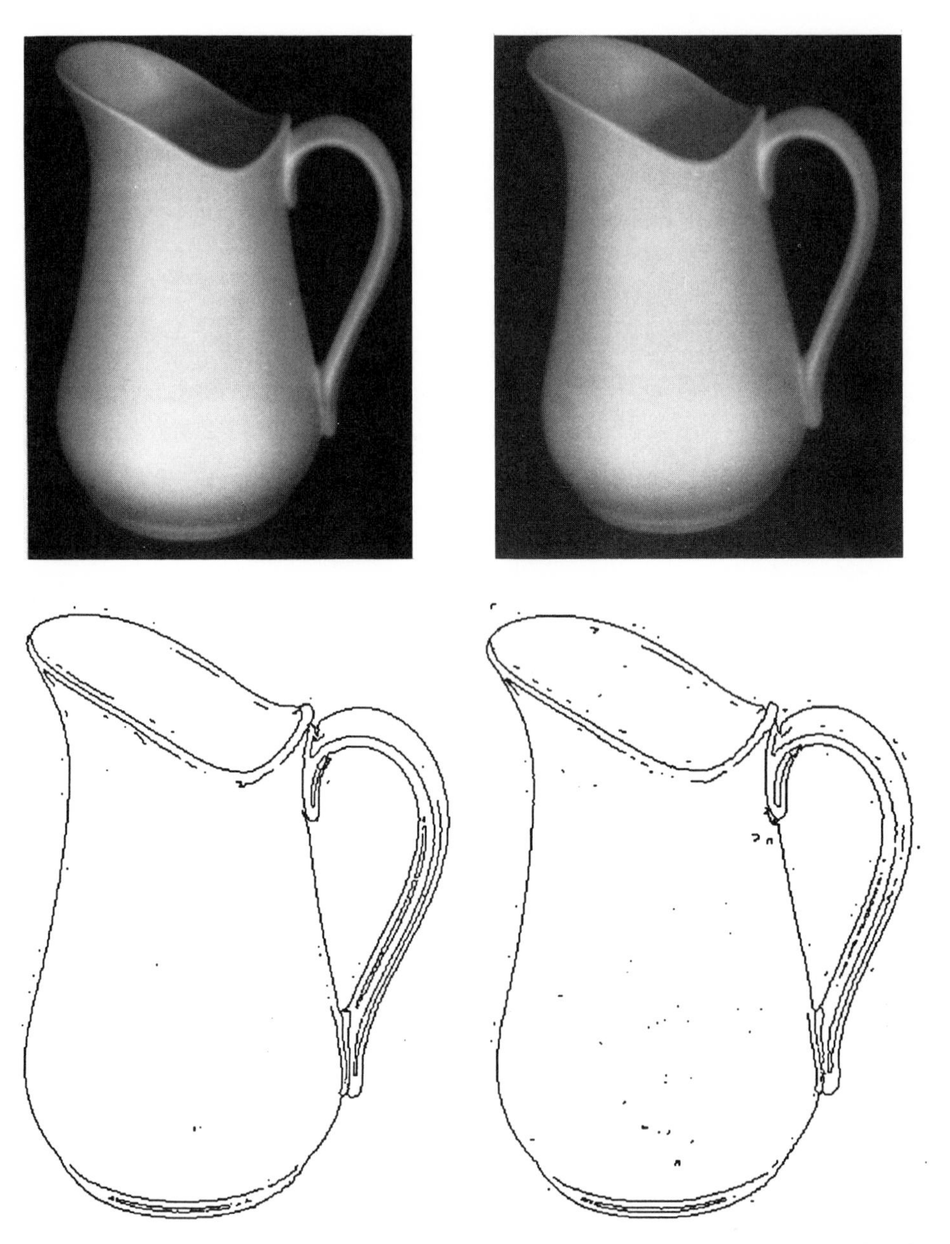

Figure 8.19. Digital images and zero-crossings for the pitcher, side view.

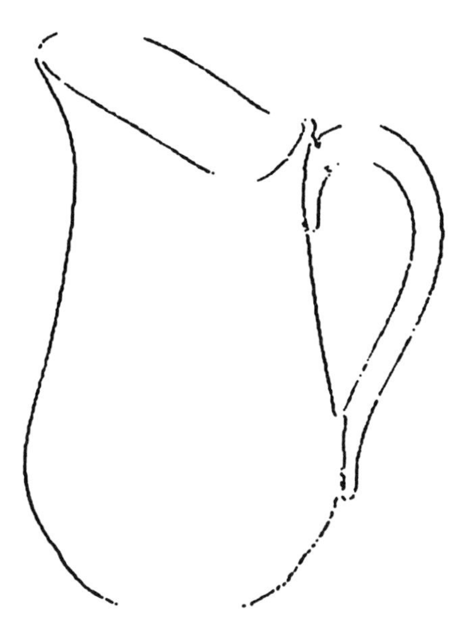

Figure 8.20. Stereo matches for the pitcher, side view.

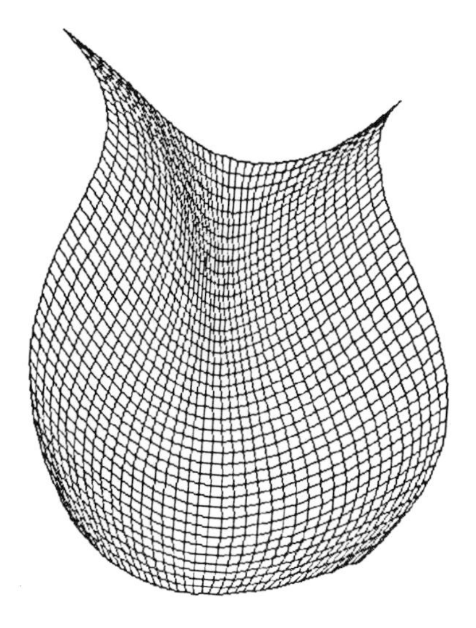

Figure 8.21. Level 1 surface for the pitcher body, side view.

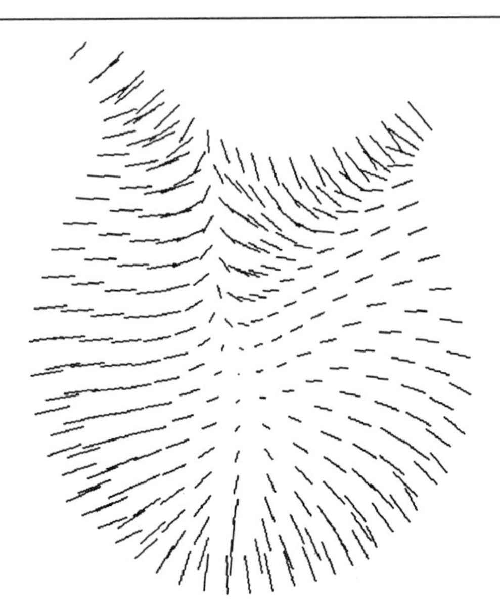

Figure 8.22. Surface normals for the pitcher body, side view.

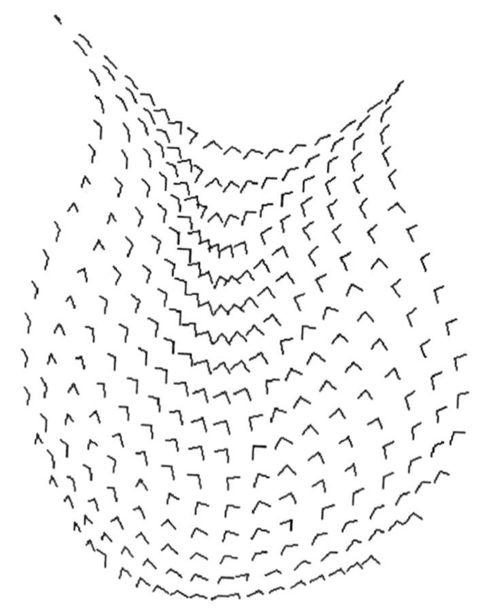

Figure 8.23. Principal directions, pitcher body, side view.

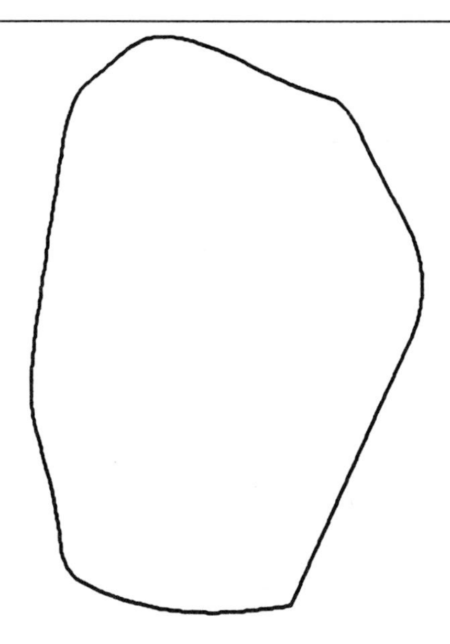

Figure 8.24. Smoothed boundary curve for pitcher hole.

Surface Analysis, pitcher body				
Surface	Area	K_{max}	K_{min}	r
sensed	26469	0.000841	-0.001439	16.335
model	59427	0.001908	-0.004580	40.67

Table 8.8. Surface analysis of the sensed and model pitcher bodies.

Moments, pitcher hole				
cavity	M_{00}	M_{20}	M_{02}	$M_{20}+M_{02}$
sensed	2353	343564	611903	955467
model	2565	1083210	283496	1366706
scaled sensed	2565	408236	727086	1135322

Table 8.9. Moments for sensed and model pitcher hole.

8.7. EXPERIMENT 6

In this experiment the pitcher was imaged from the front, with no cavity or hole in the scene. The digital images and zero-crossings are in figure 8.25 and th stereo matches are in figure 8.26 . The single region of the pitcher was traced with the sensor under active control of the system and the surface that was built is shown in figure 8.27 . Figure 8.28 shows the directions of the surface normals on the patch and figure 8.29 shows the principal directions on the surface. These are the directions of maximum and minimum curvature on the surface, and define an intrinsic coordinate system on the surface. Table 8.10 contains the surface analysis, which shows that the surface has negative Gaussian curvature which precludes it from being cylindrical or planar.

Matching of this surface was much more difficult than any of the others and was not entirely successful. The only surface of equivalent area and surface type was the pitcher. However, computing the transformation from model to sensed coordinates was not successful. Unlike a planar or cylindrical surface, an arbitrary curved surface has fewer invariants such as the normal to a planar surface or a cylindrical axis that can be matched against. Compounding this problem is the surface subset problem: even though we may match correctly, the transformation may not be easily computed since it is unclear which subset of the surface is matching which part of the larger surface.

The result of this experiment is that there is too little information to effectively know the object's structure. Therefore a new visual view is needed and this can be reported to the camera system. A slight change in viewing angle will reveal the cavity or the hole which can be used to compute the transformation matrix. Even though full recognition was not accomplished in this case, the ability to do partial matching is an improvement over vision systems that must have a global match or none at all. The discovery and quantification of a 3-D surface is useful and as the new view is taken, this information can be used to build on the description.

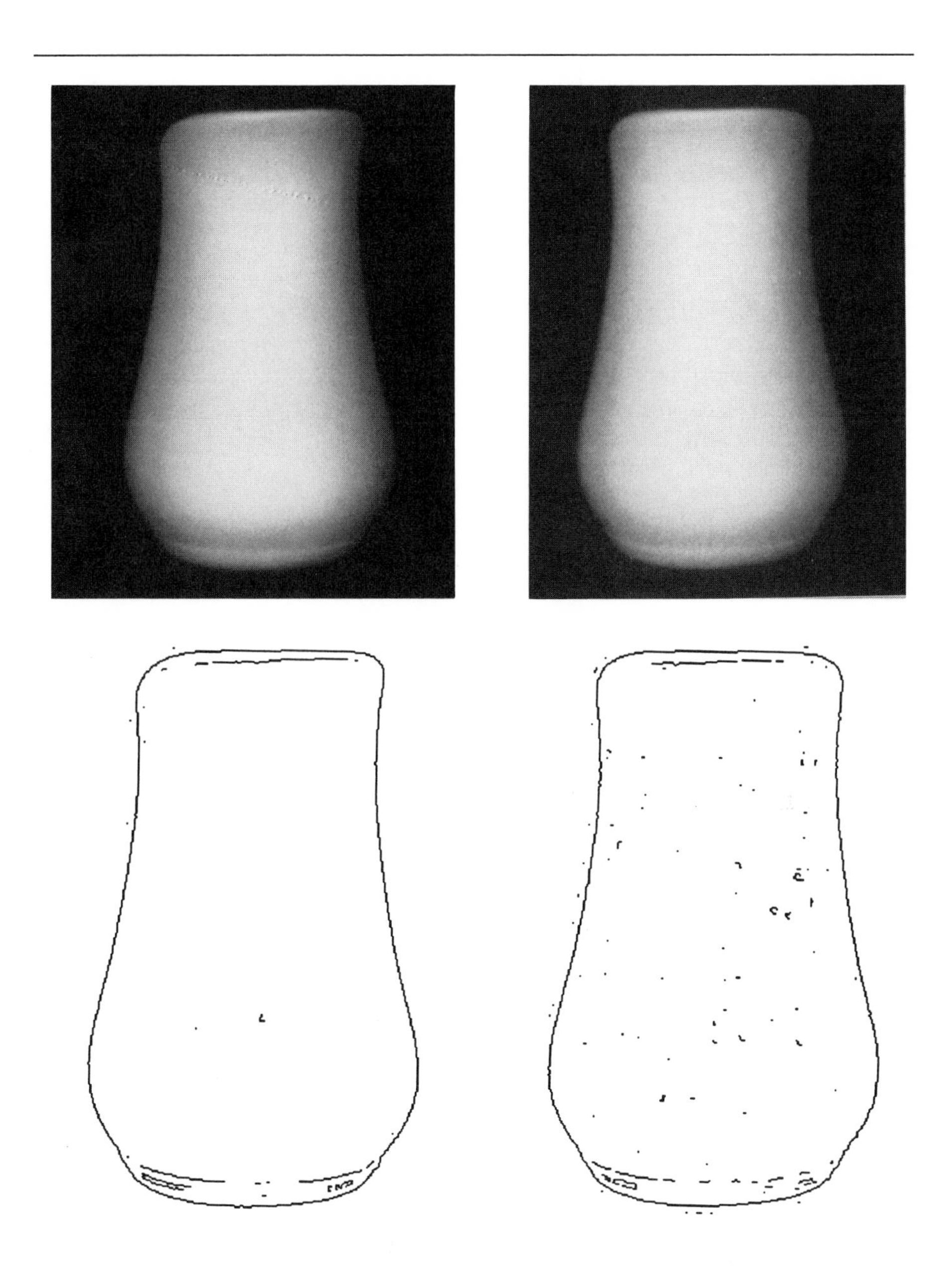

Figure 8.25. Digital images, zero-crossings for the pitcher, front view.

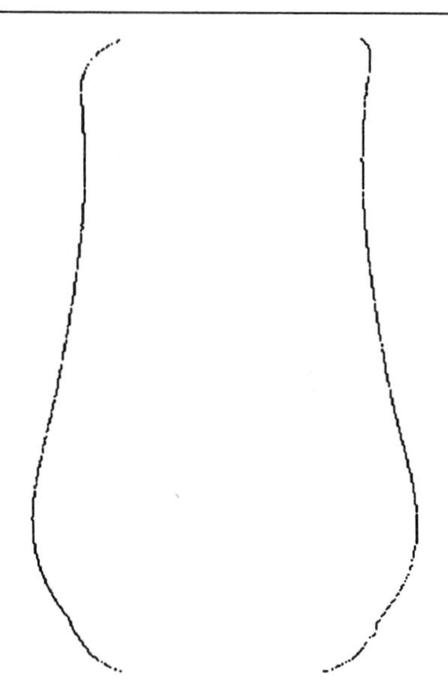

Figure 8.26. Stereo matches for the pitcher, front view.

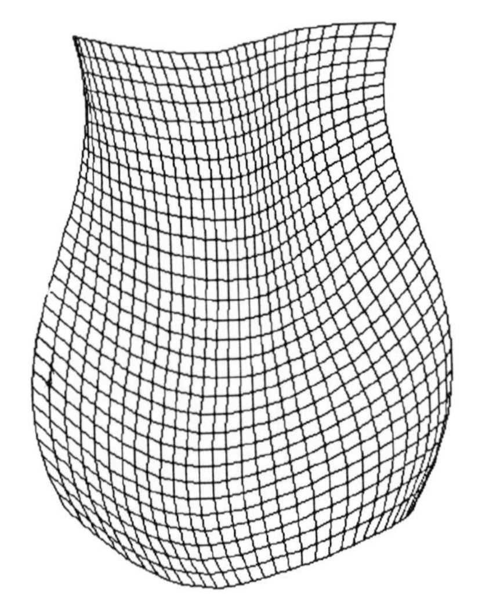

Figure 8.27. Level 1 surface for the pitcher body, front view.

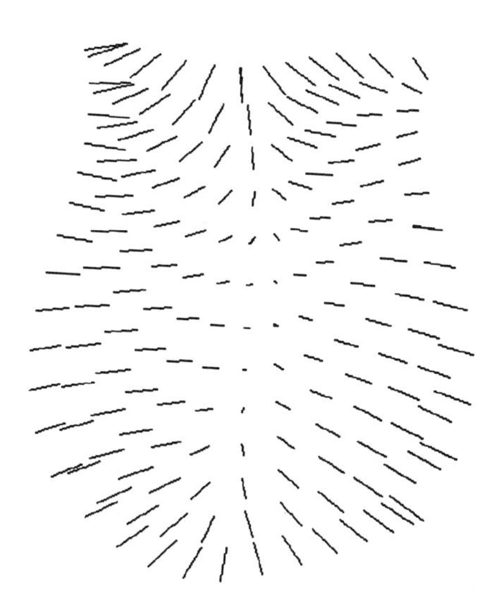

Figure 8.28. Surface normals for the pitcher body, front view.

Figure 8.29. Principal directions, pitcher body, front view.

Surface Analysis, Pitcher body, front view.				
Surface	Area	K_{max}	K_{min}	r
sensed	26747	0.000962	-0.001033	16.77
model	59427	0.001908	-0.004580	40.67

Table 8.10. Surface analysis of the sensed and model pitcher bodies.

8.8. EXPERIMENT 7

In this experiment, the coffee mug was imaged with the handle occluded. The digital images and zero-crossings are in figure 8.30 and the stereo matches in figure 8.31. The first region probed by tactile sensing was the cavity and the moment set in table 8.11 was computed. The second region probed was the body of the mug and a level 1 surface description was computed from vision and touch, shown in figure 8.32 along with the surface normals in figure 8.33. The objects in the data base that match with these two regions are a drinking glass without a handle and a mug with a handle. From this visual angle there is no way that the two objects can be distinguished. The instantiation module will pick both objects to be verified. The matcher will order these objects by complexity, causing the coffee mug to be verified. The mug is the more complex object and its only unfilled slots are the handle, hole and bottom surface. All regions from the vision analysis are matched, leaving visually occluded parts only.

It is possible to reason about and sense occluded features. It can be determined that the object is a mug by verifying the occluded hole. From the analysis so far there is no way to determine where the hole lies. If it is a mug, the hole lies in the occluded area which is shown in figure 8.34 . The bounds on this volume are known from the vision and touch sensing that has already been performed. The cavity and the hole each have an internal frame associated with them. In a rigid object, once these frames are defined, then knowing one frame determines the other through a series of transformations described in chapter 7. The object can be identified uniquely if the cavity has a unique internal frame. Knowing this internal frame and the relative transform from the model frame to the hole will allow us to compute the hole frame in the sensed coordinate system. The cavity does not possess a unique frame; it is rotationally symmetric, leaving a degree of freedom in its internal frame which is the rotation about its approach axis. This degree of freedom can be exploited to reason about the hole. The fixing of the cavity's approach axis in space means that the hole centroid is confined to lie in a circle centered at the cavity and swept out about the cavity's axis. Computing this circle gives a set of 3-D points which represent possible locations of the hole's centroid. Intersecting this circle with the known occluded volume yields a possible set of locations of the hole. Each of these locations is associated with a particular fixing of the rotationally symmetric axes about the cavity's axis. The approach is to fix the cavity's rotationally symmetric axes at an angle of rotation that is midway between the angles that bring the hole into occlusion and bring it out. Once defined, this yields an approach axis for the hole which the sensor can then use to actually sense the hole. In the experiment, the hole was found this way, rejecting the drinking glass match and accepting the mug match. Figure 8.35 shows the sensor searching for and finding the hole in the visually occluded area.

This last experiment shows the effectiveness of this approach to object recognition. Multiple sensors were used synergistically to invoke a possible set of objects. High level reasoning about the object's structure that is encoded in 3-D models allowed further verification sensing to successfully discriminate the objects. The knowledge about the 3-D world (the occluded volume) and the object's geometry (which is encoded in the model) can be used to perform active sensing in occluded areas.

8.9. SUMMARY

The experiments reported here show the ability of vision and touch sensing to sense and recognize objects that would be difficult for vision alone. The 3-D surface and feature primitives provide strong matching criteria that can lead to unique instantiations based upon a combination of surface and feature attributes. In cases of multiple consistent objects, verification sensing and high level reasoning can discriminate by sensing occluded areas. In some cases the system is unable to accurately sense features due to the sensor's poor spatial resolution and physical size. The feature matching provides strong discriminating evidence in choosing a possible object. The surface information is also able to constrain the set of possible matches. The combination of both provides strong recognition criteria.

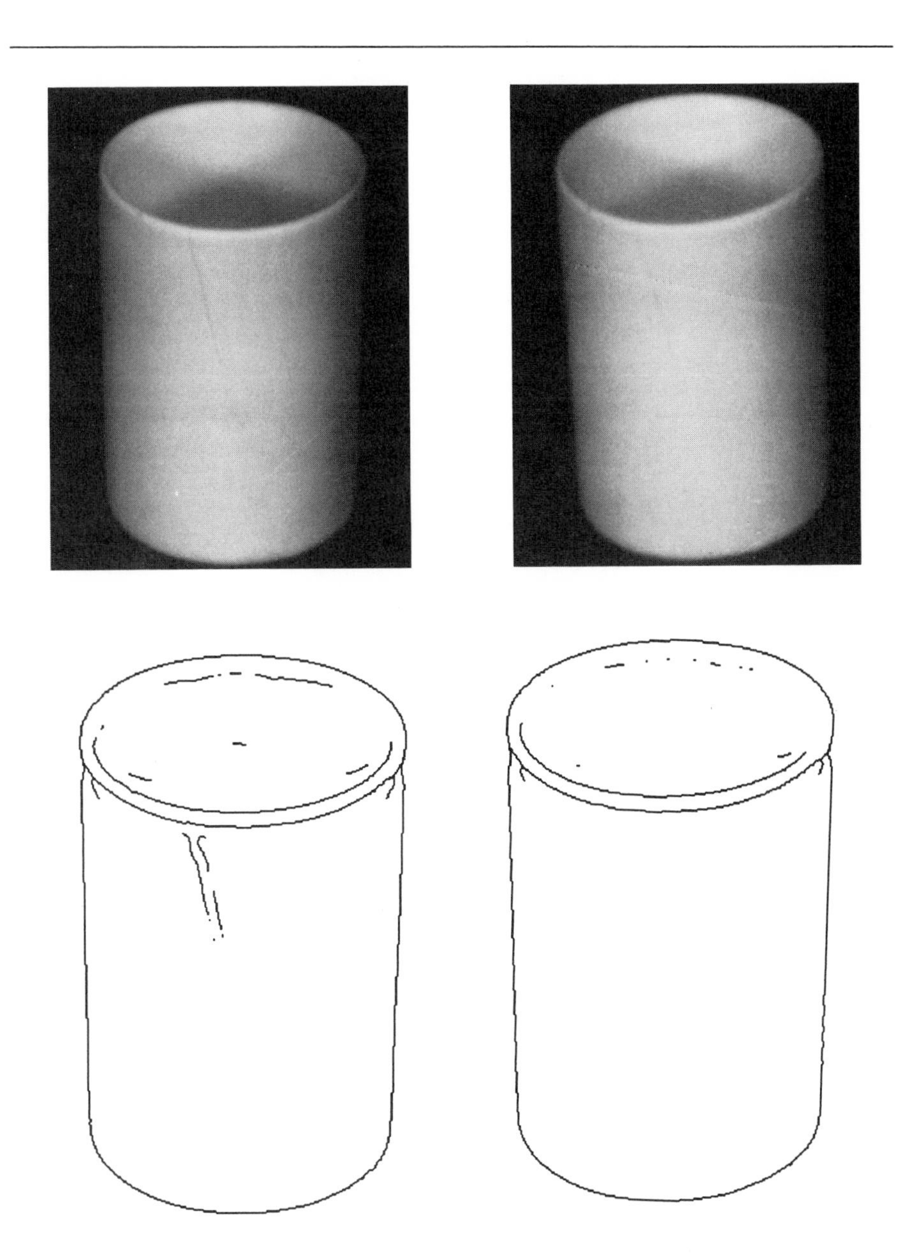

Figure 8.30. Digital images and zero-crossings for coffee mug.

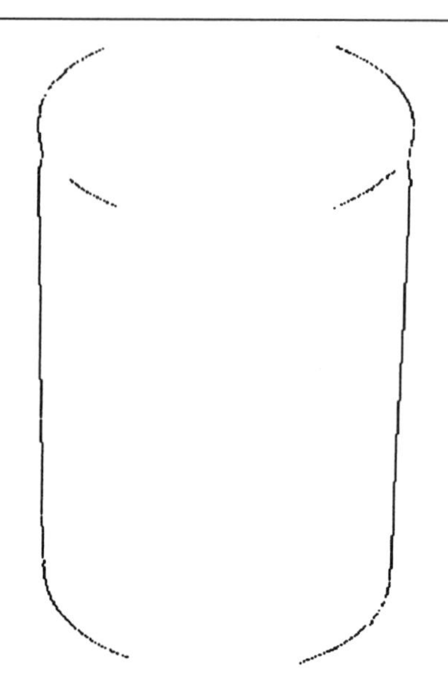

Figure 8.31. Stereo matches for the coffee mug.

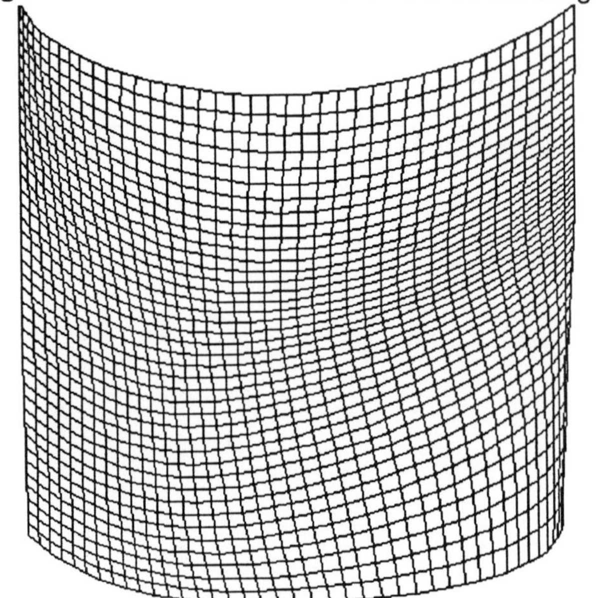

Figure 8.32. Level 1 surface for the coffee mug body.

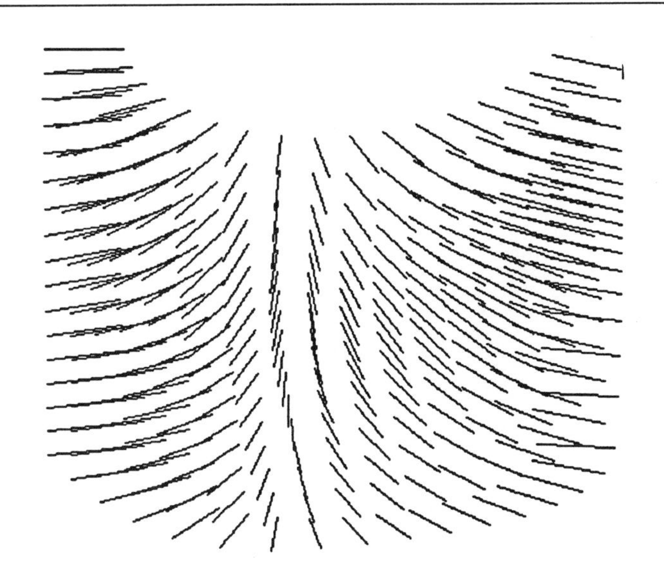

Figure 8.33. Surface normals for the coffee mug body.

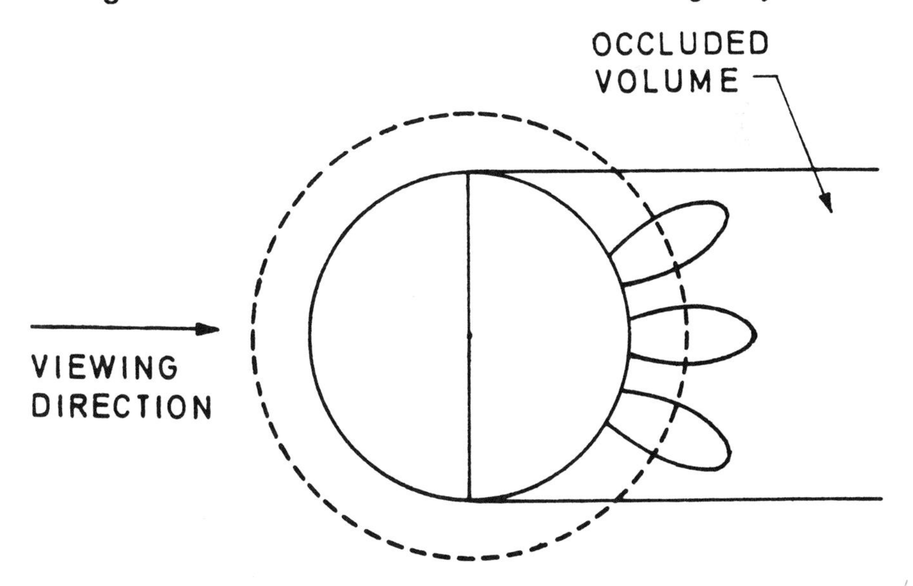

Figure 8.34. Occluded area of the coffee mug.

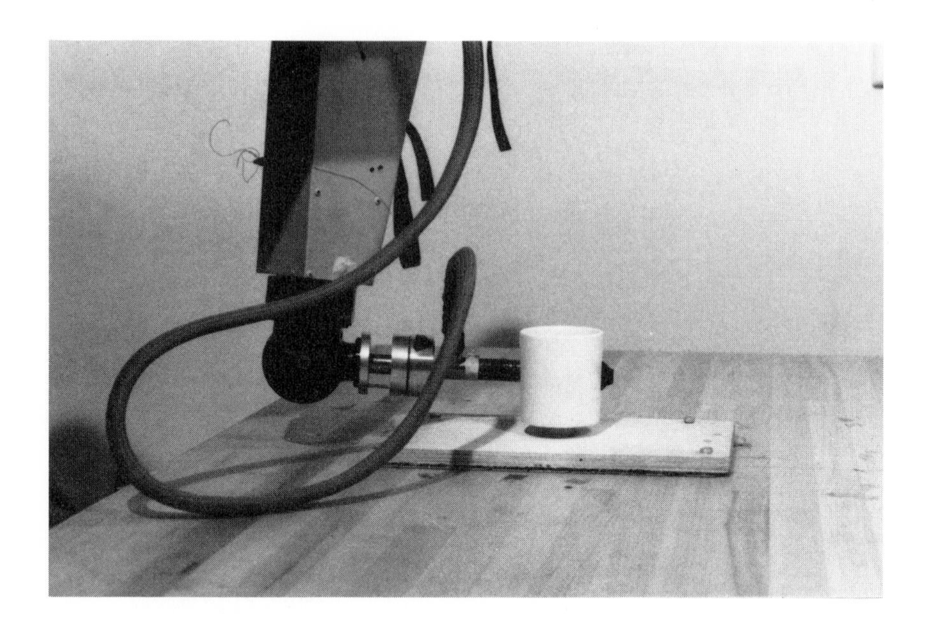

Figure 8.35. Verifying the visually occluded hole.

Moments, coffee mug cavity				
cavity	M_{00}	M_{20}	M_{02}	$M_{20}+M_{02}$
sensed	4948	1872094	2114266	3986360
model	4758	1802083	1802083	3604167
sensed scaled	4758	1731206	1955153	3686359

Table 8.11. Moments for sensed and model coffee mug cavity.

CHAPTER 9

CONCLUSION

9.1. INTRODUCTION

The two central ideas that the theoretical background and experimental results described in the previous chapters expound are the importance of discovering and understanding 3-D structure of objects and the multi-sensor approach to robotics. This chapter discusses these two ideas and outlines limitations of the system and possible future directions for this research

9.2. DISCOVERING 3-D STRUCTURE

Object recognition in this work is predicated upon discovering 3-D structure of objects. It may seem obvious that understanding 3-D structure is a necessary first step to a host of important robotic tasks, including recognition, grasping, manipulation assembly and inspection. However, this has not been the case for many previous recognition systems. Instead of being the primary initial focus, 3-D structure was an outcome of the model matching phase. Only by correctly invoking a model (determined through a variety of viewpoint-dependent, 2-D projective methods) was the actual 3-D structure uncovered. By using active sensors, 3-D structure can be discovered initially. The reasons why this is important are listed below:

- The sensed primitives need to be related to the model components in model-based recognition. The models can be easily and efficiently structured as 3-D surfaces and features. The discovery of 3-D surfaces and features facilitates this matching effort. The models in this work use the same surface primitive that the sensors compute. This eliminates expensive transformations of the data and possible information loss.
- Viewpoint-independent recognition assumes no characteristic views of the object. The orientation in space of the object needs to be computed from the combination of sensing and high level reasoning. Uncovering the 3-D structure makes this computation possible.
- There is a limit to the amount of recognition that can be done at the low level. Reasoning about 3-D objects at a higher level implies understanding the 3-D structure and creating 3-D primitives.
- Tasks beyond recognition also imply an understanding of 3-D structure. Grasping, manipulation, assembly and inspection all involve understanding and reasoning about 3-D structure.

A key component to discovering and understanding 3-D structure is the high level knowledge built into the system. Both the object models and the matching rules are needed to place the low level sensor data into context.

9.3. THE MULTI-SENSOR APPROACH

The experiments in chapter 8 showed that two sensors are better than one. A multi-sensor approach to robotics is important for a number of reasons. First, multi-sensors provide more robust and stable sensing primitives. Neither of the sensors alone are capable of building true 3-D primitives that are more than point based. Point-based methods tend to be weak and intolerant of error; relying on a single pixel or point contact is inherently unstable. The data needs to be abstracted and smoothed into larger more robust measures. In this work, the stereo matches are pixel based but the pixels to be matched are found on contours of related pixel chains, thus eliminating many possible spurious point matches. The 3-D data from these matches is not used as isolated sets of points, but grouped into curves in space, broken at discontinuities in curvature. The same is true for the tactile trace points, which are grouped into boundary curves that generate surfaces. Surfaces are large, stable entities as opposed to points. Small local discontinuities due to sensor error are reduced with this method.

Secondly, the complementary nature of the sensors allows them to accurately sense 3-D structure. What makes tactile sensing succeed in this work is the cues that vision provides; where to sense, at what orientation, in what direction. Without this higher level guidance, the touch is too difficult to control and the signals too conflicting to succeed. When a surface is traced, the sensor knows the smooth nature of the surface from the vision analysis. These constraints are combined to allow touch to actively explore surfaces and features.

Third, one sensor can compensate for anothers deficiencies. A clear advantage of tactile sensing over vision is that touch can deal with occlusion. As experiment 7 (finding the occluded mug handle) showed, touch can verify visually obscured parts of the object. It was only able to do this by using higher level reasoning about the object's structure and guide the tactile sensor to a probable location for the hole. A strategy of sending the sensor into the occluded area and reporting back contacts would be doomed to failure. The knowledge to interpret the contacts must be available.

9.4. LIMITATIONS OF THE SYSTEM

The methods described in this work will not work for small surface structures. As objects become more complex, these structures occur more often. Part of this problem is due to sensor resolution. More powerful and higher resolution sensors can conceivably deal with this problem, such as the inability to sense the handle of the pitcher or mug. However, as objects become more complex, so do the object models that define them. The models used in this work are organized by an obvious object structure and segmentation. More complex objects may not show this structure, and the models may prove inadequate. While this is a serious problem, it still does not preclude partial matching which is one of the obvious strengths of this method.

Another limitation with the system is that because it relies partially on vision, it cannot be isolated from the unsolved problems of machine vision. The creation of closed contours from edge detection is still a problem in certain scenes. A partial solution was implemented in this work but some of the images still needed a small number of pixels filled in to create the contours.

Certain degenerate viewing points of the object yield confusing sets of information that can cause problems with the tactile sensing. Typically these can be seen by noticing small structure and responding to it, yielding partial matches as discussed above. However, the higher level reasoning modules are not yet developed to try to reason about this.

Homogeneous objects were used with a purposeful lack of detail. However, in real world robotics, objects will have structure encompassing texture, reflectivity changes on the surface and noisy surface gradients. (Think of your favorite mug with a design on its surface that will totally confuse vision.) Tactile sensing can help in this case since surface structure can be sensed independent of printing on the structure.

9.5. FUTURE DIRECTIONS

There are a number of directions in which this research leads. This work has shown the utility of sensing to reason about and understand 3-D structure. The sensing that is done is done serially, first using vision and then touch to create a hypothesis and then using touch to verify. More tactile feedback could be used to help support or reject a visual hypothesis. Vision is relied upon to create regions of interest. Tactile sensing could in fact verify that these regions really are regions that have physical meaning before a detailed surface or feature trace is attempted. Further, vision could reevaluate what it knows about the image based on the tactile discovery. Many vision algorithms show markedly improved performance when they are given constraints on a scene, such as a region's being cylindrical or planar. The tactile can only make the vision more robust as it determines 3-D structure.

The data base of models includes single objects. Extending this to multiple objects in a scene and articulated parts would be a useful extension. Both of these ideas could be implemented with the coordinate frame method, where object relationships are described in terms of relative frames with variables instead of precomputed constants.

The part of this work that is the least well developed is the higher level reasoning about solid objects. This ability is clearly a must for robotics to succeed in the future. It requires efforts in 1) building automatic 3-D models of objects that capture the object's complexity and 2) relating the spatial information contained in the model to goals and further sensing.

Finally, the sensor used here was a single finger. Robots will need multiple fingers to perform grasping and manipulation. Extending this work to multiple fingers is possible. One method would be to use the other fingers to try to stabilize the object during tactile sensing; the objects used now are rigidly attached to the support surface. The other would be to have multiple parallel traces implemented, for which there may be psychological evidence showing that this is an effective human strategy also. Multi-fingered robotic hands are potentially capable of extremely complex recognition tasks and their development will go far in fulfilling the promise of robotics.

References

1. Agin, G., "Representation and description of curved objects," *Stanford University A.I. Memo*, no. 173, October 1972.
2. Allen, Peter and Ruzena Bajcsy, "Converging disparate sensory data," *Proc. 2nd International Symposium on Robotics Research*, Kyoto, August 1984.
3. Badler, N. and R. Bajcsy, "Three dimensional representations for computer graphics and computer vision," *Proc. SIGGRAPH 78*, vol. 12, no. 3, pp. 153-160, August 1978.
4. Bajcsy, R. and L. Lieberman, "Texture gradient as a depth cue," *Computer Graphics and Image Processing*, vol. 5, no. 1, pp. 52-67, March 1976.
5. Barnard, Stephen and Martin Fischler, "Computational stereo," *ACM Computing Surveys*, vol. 14, no. 4, pp. 553-572, December 1982.
6. Bhanu, B. and O. Faugeras, "Shape matching for two-dimensional objects," *IEEE trans. on Pattern Analysis and Machine Intelligence*, vol. PAMI-6, no. 2, pp. 137-155, March 1984.
7. Binford, T., "Survey of model based image analysis," *Int. Journal of Robotics Research*, vol. 1, no. 1, pp. 18-64, Spring 1982.
8. Boas, Mary L., *Mathematical methods in the physical sciences,* John Wiley, New York, 1983.
9. Bolles, R. C., P. Horaud, and M. J. Hannah, "3DPO: A three dimensional part orientation system," *Proc. 8th IJCAI*, Karlsruhe, Germany, August 1983.
10. Boyse, J. W., "Interference detection among solids and surfaces," *CACM*, vol. 22, pp. 3-9, January 1979.
11. Brady, Michael and Haruo Asada, "The curvature primal sketch," A.I. memo 758, M.I.T A.I. Laboratory, Cambridge, February 1984.
12. Brady, Michael, Jean Ponce, Alan Yuille, and Haruo Asada, "Describing surfaces," Joint U.S. France NSF-CNRS Workshop on Robotics, Philadelphia, November 7-9, 1984.
13. Brooks, Rodney, "Symbolic reasoning among 3-D models and 2-D images," *Artificial Intelligence*, vol. 17, pp. 285-349, 1981.

14. Clocksin, W. F. and C. S. Mellish, *Programming in Prolog,* Springer Verlag, Berlin, 1981.

15. Faux, I. D. and M. J. Pratt, *Computational geometry for design and manufacture,* John Wiley, New York, 1979.

16. Fisher, R. B., "Using surfaces and object models to recognize partially obscured objects," *Proc. IJCAI 83*, pp. 989-995, Karlsruhe, August 1983.

17. Foley, James and Andries Van Dam, *Fundamentals of interactive computer graphics,* Addison-Wesley, Reading, 1982.

18. Forrest, A. R., "On Coons' and other methods for the representations of curved surfaces," *Computer Graphics and Image Processing,* vol. 1, pp. 341-359, 1972.

19. Ganapathy, S., "Decomposition of transformation matrices for robot vision," *Proc. IEEE International Conference on Robotics*, pp. 130-139, Atlanta, March 13-15, 1984.

20. Gordon, G., *Active touch, the mechanism of recognition of objects by manipulation,* Pergamon Press, 1978.

21. Grimson, W. E. L., *From images to surfaces: A computational study of the human early visual system,* MIT Press, Cambridge, 1981.

22. Grimson, W. E. L. and Tomas Lozano-Perez, "Model based recognition and localization from sparse three dimensional sensory data," A.I. memo 738, M.I.T. A.I. Laboratory, Cambridge, August 1983.

23. Harmon, Leon, *Touch sensing technology,* Technical Report MSR80-03, Society of Manufacturing Engineers, Dearborn , 1980.

24. Harmon, Leon, "Automated tactile sensing," *Int. Journal of Robotics Research*, vol. 1, no. 2, pp. 3-32, Summer 1982.

25. Henderson, Thomas C., "Efficient 3-D object representations for industrial vision systems," *IEEE Transactions on Pattern Analysis and Machine Intelligence*, vol. PAMI-5, no. 6, November 1983.

26. Hilbert, D. and S. Cohn-Vossen, *Geometry and the imagination,* Chelsea, New York, 1952.

27. Hillis, W. D., "A high resolution imaging touch sensor," *Int. Journal of Robotics Research*, vol. 1, no. 2, pp. 33-44, Summer 1982.

28. Hoffman, D., "The interpretation of visual illusions," *Scientific American*, pp. 154-162, November 1983.

29. Horn, B. K. P., R. Woodham, and W. M. Silver, "Determining shape and reflectance using multiple images," *AI memo 490*, MIT AI Laboratory, Cambridge, 1978.

30. Hu, Ming-Kuei, "Visual pattern recogniton by moment invariants," *IEEE Transactions on Information Theory*, vol. IT-8, pp. 179-187, February 1962.

31. Izaguirre, A., "Implementing remote control of a robot using the VAL-II language," Technical Report , Grasp Laboratory, University of Pennsylvania, Philadelphia, 1984.

32. Izaguirre, A., P. Pu, and J. Summers, "A new development in camera calibration calibrating a pair of mobile cameras," *Proc. IEEE International Conference on Robotics and Automation*, pp. 74-79, St. Louis, March 25-28, 1985.

33. Kalvin, Alan, Jacob T. Schwartz, Edith Schonberg, and Micha Sharir, "Two dimensional model based boundary matching using footprints," Technical Report, Courant Institute, NYU, New York, 1985.

34. Kender, J. R., "Shape from texture: a brief overview and a new aggregation transform," *Proc. DARPA IU Workshop*, pp. 79-84, November 1978.

35. Kinoshita, G., S. Aida, and M. Mori, "A pattern classification by dynamic tactile sense information processing," *Pattern Recognition*, vol. 7, pp. 243-250, 1975.

36. Koenderink, J. J. and A. J. van Doorn, "The internal representation of solid shape with respect to vision," *Biological Cybernetics*, vol. 32, pp. 211-216, 1979.

37. Koparkar, P. A. and S. P. Mudur, "Computational techniques for processing parametric surfaces," *Computer Vision, Graphics and Image Processing*, vol. 28, no. 3, December 1984.

38. Kuan, Darwin and Robert Drazovich, "Intelligent interpretation of 3-D imagery," *AI&DS technical report*, vol. 1027-1, AI&DS, Mountain View, 1983.

39. Marr, David, *Vision,* W. Freeman, San Francisco, 1982.

40. Marr, David and Ellen Hildreth, "Theory of edge detection," *Proc. Royal Society of London Bulletin*, vol. 204, pp. 301-328, 1979.

41. Marr, David and Tomaso Poggio, "Cooperative computation of stereo disparity," *Science*, vol. 194, pp. 283-287, 1976.

42. Nevatia, R. and K. R. Babu, "Linear feature extractor and description," *Computer Graphics and Image Processing*, vol. 13, pp. 257-269, 1980.

43. Nevatia, R. and T. Binford, "Description and recognition of curved objects," *Artificial Intelligence*, vol. 8, pp. 77-98, 1977.

44. Nitzan, D., "Assessment of robotic sensors," *Proc. 1st International Conference on Robot Vision and Sensory Controls*, Stratford-upon-Avon, UK, April 1-3, 1981.

45. Okada, T. and S. Tsuchiya, "Object recognition by grasping," *Pattern Recognition*, vol. 9, pp. 111-119, 1977.

46. Oshima, M. and Y. Shirai, "Object recognition using three dimensional information," *Pattern Recognition*, vol. 11, pp. 9-17, 1979.

47. Oshima, M. and Y. Shirai, "Object recognition using three dimensional information," *IEEE trans. on Pattern Analysis and Machine Intelligence*, vol. PAMI-5, no. 4, pp. 353-361, July 1983.

48. Overton, K. J., "The acquisition, processing and use of tactile sensor data in robot control," Ph.D. Dissertation, University of Massachusetts, Amherst, May 1984.

49. Ozaki, H., S. Waku, A. Mohri, and M. Takata, "Pattern recognition of a grasped object by unit vector distribution," *IEEE trans. n Systems, Man and Cybernetics*, vol. SMC-12, no. 3, pp. 315-324, May/June 1982.

50. Pavlidis, T., *Algorithms for graphics and image processing,* Computer Science Press, Rockville, MD, 1982.

51. Potmesil, Michael, "Generating three dimensional surface models of solid objects from multiple projections," IPL technical report 033, Image Processing Laboratory, RPI, Rensselaer, October 1982.

52. Requicha, A. A. G., "Representations for rigid solids: theory, methods and systems," *ACM Computing Surveys*, vol. 12, no. 4, December 1980.

53. Rock, Irvin, *The logic of perception,* MIT Press, Cambridge, 1983.

54. Rosenfeld, Azriel and A. Kak, *Digital Picture Processing, 2nd edition,* Academic Press, New York, 1982.

55. Rosenfeld, A. and Emily Johnston, "Angle detection on digital curves," *IEEE Transactions on Computers*, vol. C-22, pp. 875-878, 1973.

56. Selesnick, S. A., "Local invariants and twist vectors in computer-aided geometric design," *Computer Graphics and Image Processing*, vol. 17, no. 2, pp. 145-160, October 1981.

57. Shapiro, Linda and Robert Haralick, "A hierarchical relational model for automated inspection tasks," *Proc. IEEE International Conference on Robotics*, pp. 70-77, Atlanta, March 13-15, 1984.

58. Shapiro, Linda, J. D. Moriarty, R. Haralick, and P. Mulgaonkar, "Matching three dimensional models," *Proc. of IEEE conference on pattern recognition and image processing*, pp. 534-541, Dallas, August 1981.

59. Shneier, M., S. Nagalia, J. Albus, and R. Haar, "Visual feedback for Robot Control," *IEEE Workshop on Industrial Applications of Industrial Vision*, pp. 232-236., May 1982.

60. Smitley, David, "The design and analysis of a stereo vision algorithm," M.S. thesis, University of Pennsylvania, Philadelphia, May 1985.

61. Solina, Franc, "Errors in stereo due to quantization," GRASP lab technical report, University of Pennsylvania, Philadelphia, July 1985.

62. Stevens, Kent, "The visual interpretation of surface contours," *Artificial Intelligence*, vol. 17, pp. 47-75, 1981.

63. Sutherland, I. E., "Three dimensional input by tablet," *Proc. of IEEE*, no. 62 (4), pp. 453-461, April 1974.

64. Tomita, Fumiaki and Takeo Kanade, "A 3D vision system: Generating and matching shape descriptions in range images," *IEEE conference on Artificial Intelligence Applications*, pp. 186-191, Denver, December 5-7, 1984.

65. Unimation,, *PUMA Mark II robot equipment manual,* Unimation Inc., Danbury, February 1983.

66. Unimation,, *User's guide to VAL-II,* Unimation Inc., Danbury, April 1983.

67. Witkin, Andrew, "Recovering surface shape and orientation from texture," *Artificial Intelligence*, vol. 17, pp. 17-47, 1981.

68. Witkin, Andrew, "Scale-Space filtering," *Proc. 8th IJCAI*, pp. 1019-1022, Karlsruhe, Germany, August 1983.

69. York, B., "Shape representation in computer vision," Ph.D. dissertation, University of Massachusetts, Amherst, 1981.

APPENDIX

BICUBIC SPLINE SURFACES

1. INTRODUCTION

The surfaces that are used in modeling the objects are represented as parametric bicubic surface patches. The integration of vision and touch to build surface descriptions also uses this representation. Therefore it is instructive to explore this representation fully. Faux and Pratt [15], Foley and Van Dam [17] and Forrest [18] contain a more detailed discussion of bicubic patches, and this appendix draws from these references.

2. PARAMETRIC CURVES AND SURFACES

The parametric form of a space curve $\mathbf{P}(u)$ parameterized by u is:

$$\mathbf{P}(u) = (x(u), y(u), z(u)) \tag{A.1}$$

This representation is not unique, as there are a number of different parameterizations that yield the same curve. The tangent vector of a parametric curve **P** is:

$$\mathbf{P}_u(u) = (\frac{dx}{du}, \frac{dy}{du}, \frac{dz}{du}) \tag{A.2}$$

For a surface, two parameters are needed. The parametric representation is:

$$\mathbf{P}(u,v) = (x(u,v), y(u,v), z(u,v)) \tag{A.3}$$

The tangents in each of the parametric directions on the surface are:

$$\mathbf{P}_u(u,v) = (\frac{\partial x}{\partial u}, \frac{\partial y}{\partial u}, \frac{\partial z}{\partial u}) \tag{A.4}$$

$$\mathbf{P}_v(u,v) = (\frac{\partial x}{\partial v}, \frac{\partial y}{\partial v}, \frac{\partial z}{\partial v}) \tag{A.5}$$

The unit surface normal $\mathbf{n}(u,v)$ at a point on the surface is formed by taking the cross product of the the tangents in each of the parametric directions:

$$\mathbf{n} = \frac{\frac{\partial \mathbf{P}}{\partial u} \times \frac{\partial \mathbf{P}}{\partial v}}{\left| \frac{\partial \mathbf{P}}{\partial u} \times \frac{\partial \mathbf{P}}{\partial v} \right|} \tag{A.6}$$

3. COONS' PATCHES

The particular form of bicubic surface patch that is being used in this research was originally studied by S.A. Coons and is known as a Coons' patch. Coons' formulation of this type of surface patch was somewhat more general and the restricted form of Coons' patch used here is sometimes referred to as a tensor product, Cartesian product or Ferguson surface. These patches have been used extensively in computer graphics and computer aided design. The patches are constructive in that they are built up from known data and are interpolants of sets of three dimensional data defined on a rectangular parametric mesh. This gives them the advantage of axis independence, which is important in both modeling and synthesizing these patches from sensory data.

3.1. LINEARLY INTERPOLATED PATCHES

The surface interpolation problem that is being considered here is to define a mapping from the unit parametric square plane into a surface defined on R^3:

$$P : [0,1] \times [0,1] \rightarrow R^3 \tag{A.7}$$

such that the mapping interpolates the data points specified. To create such a mapping, we choose four points

$$\mathbf{P}(0,0)\ ,\ \mathbf{P}(0,1)\ ,\ \mathbf{P}(1,0)\ ,\ \mathbf{P}(1,1) \tag{A.8}$$

which form the vertices of the patch and are referred to as the knot points (figure 6.1). These points are defined on the parametric grid

$$0 \leq u\ ,v \leq 1$$

If we form line segments between adjacent knots as the bounding contours of the patch, we can create an interpolated surface patch bounded by the line segments:

$$\mathbf{P}(0,v)\ ,\ \mathbf{P}(u,0)\ ,\ \mathbf{P}(1,v)\ ,\ \mathbf{P}(u,1) \tag{A.9}$$

To interpolate the interior of this patch, we can linearly interpolate between the curves on opposite sides of the patch; between $\mathbf{P}(0,v)$ and $\mathbf{P}(1,v)$ in the u direction and similarly between $\mathbf{P}(u,0)$ and $\mathbf{P}(u,1)$ in the v direction. The equation of the surface then becomes:

$$R1 = \mathbf{P}(u,v) = \mathbf{P}(0,0)\,(1-u)\,(1-v) + \mathbf{P}(0,1)\,(1-u)\,(v) + \mathbf{P}(1,0)\,(u)\,(1-v) + \mathbf{P}(1,1)\,(u)\,(v) \tag{A.10}$$

Substituting values for u and v verifies that the boundary curves (A.9) are in fact the line segments between the knot points. This kind of a patch is referred to as a *bilinear* patch.

Having built a bilinear surface $\mathbf{P}(u,v)$ that interpolates the data points, we want to know if it is the only such surface. The answer is clearly no, as there are an infinite number of surfaces that will interpolate the sparse data at the boundaries. In constructing other surfaces, we can relax some of the above restrictions to form more complex surfaces. In particular, we need not require linear boundary curves. If we know more boundary data than just the knot points, we can form two cubic polynomial space curves $\mathbf{P}(u,0)$ and $\mathbf{P}(u,1)$ which interpolate the boundary between adjacent knots in the u direction which can then be linearly interpolated in the v direction to obtain:

$$R2 = \mathbf{P}(u,v) = \mathbf{P}(u,0)\,(1-v) + \mathbf{P}(u,1)\,(v). \tag{A.11}$$

If we know the other two boundary curves, $\mathbf{P}(0,v)$ and $\mathbf{P}(1,v)$, we can similarly form another surface:

$$R3 = \mathbf{P}(u,v) = \mathbf{P}(0,v)\,(1-u) + \mathbf{P}(1,v)\,(u). \tag{A.12}$$

$R2$ and $R3$ form ruled surfaces as they are linear in one of the parametric directions. To form a surface that has nonlinear boundary curves on all boundaries we can sum surfaces $R2$ and $R3$. However, substituting values of u and v reveals that the knot points will not be interpolated correctly nor will the boundary curves (A.9) be correct. This is due to the fact that summing these two ruled surfaces includes the corner points twice. To negate this effect, we can subtract out the unwanted terms by subtracting surface R1 to create a new surface:

$$R4 = \mathbf{P}(u,v) = R2 + R3 - R1 \tag{A.13}$$

$$\begin{aligned} &= \mathbf{P}(u,0)\,(1-v) + \mathbf{P}(u,1)\,(v) \\ &\quad + \mathbf{P}(0,v)\,(1-u) + \mathbf{P}(1,v)\,(u) \\ &\quad - \mathbf{P}(0,0)\,(1-u)\,(1-v) - \mathbf{P}(0,1)\,(1-u)\,(v) \\ &\quad - \mathbf{P}(1,0)\,(u)\,(1-v) - \mathbf{P}(1,1)\,(u)\,(v)\,. \end{aligned} \tag{A.14}$$

Substitution of u and v verifies that the knot points are correctly interpolated as are the boundary curves. This surface can also be written in matrix form as:

$$\mathbf{P}(u,v) = \begin{bmatrix} (1-u) & u \end{bmatrix} \begin{bmatrix} \mathbf{P}(0,v) \\ \mathbf{P}(1,v) \end{bmatrix} + \begin{bmatrix} \mathbf{P}(u,0) & \mathbf{P}(u,1) \end{bmatrix} \begin{bmatrix} 1-v \\ v \end{bmatrix}$$

$$- \begin{bmatrix} (1-u) & u \end{bmatrix} \begin{bmatrix} \mathbf{P}(0,0) & \mathbf{P}(0,1) \\ \mathbf{P}(1,0) & \mathbf{P}(1,1) \end{bmatrix} \begin{bmatrix} 1-v \\ v \end{bmatrix} \tag{A.15}$$

3.2. HERMITE INTERPOLATION

From the matrix representation we can see that u, $1-u$, v, $1-v$ are functions that blend together the 4 defined boundary curves and are appropriately known as *blending functions.* The blending functions in (A.15) are linear and by removing this restriction we are able to build more complex interpolating surfaces. In particular, we can specify that the blending functions be cubic polynomials, such that the knot points are still interpolated. However, by specifying boundary curve information only, we will only be able to have adjacent patches exhibit positional or C^0 continuity. Our goal is to build composite surfaces composed of many adjacent patches that have higher levels of continuity. To obtain C^1 or derivative continuity, we need to specify boundary tangent information. We must specify the positional constraint embodied in the boundary curve as well as a tangential constraint along the entire boundary curve to form a smooth join. A simple way to do this is to use Hermite interpolation between the knot points to form the boundary curves and the boundary tangent criteria. Hermite interpolation interpolates a cubic polynomial space curve between two known points, given the points and the tangents to the curve at the two points. If the curve between the two points is parameterized by u, $0 \leq u \leq 1$, then the interpolating curve $\mathbf{P}(u)$ between two points $\mathbf{P}(0)$ and $\mathbf{P}(1)$ with respective tangents $\mathbf{P}_u(0)$ and $\mathbf{P}_u(1)$ is:

$$\mathbf{P}(u) = U\ M_h\ G_h \tag{A.16}$$

$$= \begin{bmatrix} 1 & u & u^2 & u^3 \end{bmatrix} \begin{bmatrix} 1 & 0 & 0 & 0 \\ 0 & 0 & 1 & 0 \\ -3 & 3 & -2 & -1 \\ 2 & -2 & 1 & 1 \end{bmatrix} \begin{bmatrix} \mathbf{P}(0) \\ \mathbf{P}(1) \\ \mathbf{P}_u(0) \\ \mathbf{P}_u(1) \end{bmatrix} \tag{A.17}$$

where M_h is the Hermite matrix and G_h is the Hermite geometry matrix. Substitution of u=0 and u=1 shows that the endpoints and end tangents are correctly interpolated by this curve. Extending this to two dimensions, we need to specify the four boundary curves of each patch to insure positional continuity, and we also need to specify the cross boundary tangents to insure a smooth C^1 join between patches. Across the u direction boundary curves, $\mathbf{P}(u,0)$ and $\mathbf{P}(u,1)$, we need to express the tangents in the v direction and vice versa for the v direction boundary curves $\mathbf{P}(0,v)$ and $\mathbf{P}(1,v)$. Specifying these tangents can also be done by using Hermite interpolation. At each knot point we specify tangents in each of the parametric directions to create the boundary curves using the Hermite method. To create the tangent criteria along the boundary curves, we can again use the Hermite method. To create the tangent criteria along the boundary curve $\mathbf{P}(0,v)$ we must interpolate tangents in the u direction along this curve. We know the u direction tangents at the endpoints, $\mathbf{P}_u(0,0)$ and $\mathbf{P}_u(0,1)$. This gives us two of the four pieces that Hermite interpolation requires. The other two pieces are the cross derivatives at the knots. These can be thought of as the rate of change of the tangent in the v direction with respect to u, $\mathbf{P}_{uv}$ or the rate of change of the u direction tangent with respect to v, $\mathbf{P}_{vu}$, which can be shown to be equivalent [15]. The equation for a surface with these characteristics can be built analogously to (A.13). The equation simplifies below because the cubic blending functions are the same functions that are used to create the cubic Hermite boundary curves.

$$\mathbf{P}(u,v) = U\, M_h\, Q\, M_h^T\, V \tag{A.18}$$

$$= \begin{bmatrix} 1 & u & u\,2u^3 \end{bmatrix} \begin{bmatrix} 1 & 0 & 0 & 0 \\ 0 & 0 & 1 & 0 \\ -3 & 3 & -2 & -1 \\ 2 & -2 & 1 & 1 \end{bmatrix}$$

$$\begin{bmatrix} \mathbf{P}(0,0) & \mathbf{P}(0,1) & \mathbf{P}_v(0,0) & \mathbf{P}_v(0,1) \\ \mathbf{P}(1,0) & \mathbf{P}(1,1) & \mathbf{P}_v(1,0) & \mathbf{P}_v(1,1) \\ \mathbf{P}_u(0,0) & \mathbf{P}_u(0,1) & \mathbf{P}_{uv}(0,0) & \mathbf{P}_{uv}(0,1) \\ \mathbf{P}_u(1,0) & \mathbf{P}_u(1,1) & \mathbf{P}_{uv}(1,0) & \mathbf{P}_{uv}(1,1) \end{bmatrix} \begin{bmatrix} 1 & 0 & -3 & 2 \\ 0 & 0 & 3 & -2 \\ 0 & 1 & -2 & 1 \\ 0 & 0 & -1 & 1 \end{bmatrix} \begin{bmatrix} 1 \\ v \\ v^2 \\ v^3 \end{bmatrix} \tag{A.19}$$

Matrix Q above is a matrix of coefficients contained on the boundaries of the patch. The upper left 2 x 2 partition of Q is a matrix of the knot points. The upper right and lower left 2 x 2 partitions are the tangents at the knot points in each of the parametric directions. The lower right 2 x 2 partition contains the cross derivatives, known as the twist vectors, at each of the knot points. In building a composite surface with many adjoining patches, we can insure C^1 continuity across these patches by imposing the following constraints on the coefficient matrices:

> Given patch $\mathbf{P1}(u,v)$, with boundary curve $\mathbf{P1}(1,v)$, and an adjoining patch $\mathbf{P2}(u,v)$, with shared boundary curve $\mathbf{P2}(0,v)$, the rows of the coefficient matrices must conform to the following constraint:

$$Q = \begin{bmatrix} - & - & - & - \\ q_{10} & q_{11} & q_{12} & q_{13} \\ - & - & - & - \\ q_{30} & q_{31} & q_{32} & q_{33} \end{bmatrix} \quad Q2 = \begin{bmatrix} q_{10} & q_{11} & q_{12} & q_{13} \\ - & - & - & - \\ kq_{30} & kq_{31} & kq_{32} & kq_{33} \\ - & - & - & - \end{bmatrix} \qquad \text{(A.20)}$$

It can be seen that this reproduces the boundary curves on each patch and that the tangents are maintained also across the join. The constant k in (A.20) is allowed because the actual tangents are ratios of the parametric tangents, and the constant drops out when taking these ratios. Similarly, for patches joined along a u direction curve, we can replicate columns of the matrices to form a smooth join. In this case, for a surface joined along $\mathbf{P1}(u,1)$ and $\mathbf{P2}(u,0)$, the constraint is:

$$Q1 = \begin{bmatrix} - & q_{01} & - & q_{03} \\ - & q_{11} & - & q_{13} \\ - & q_{21} & - & q_{23} \\ - & q_{31} & - & q_{33} \end{bmatrix} \quad Q2 = \begin{bmatrix} q_{01} & - & kq_{03} & - \\ q_{11} & - & kq_{13} & - \\ q_{21} & - & kq_{23} & - \\ q_{31} & - & kq_{33} & - \end{bmatrix} \qquad \text{(A.21)}$$

3.3. CURVATURE CONTINUOUS PATCHES

The patches above are joined with C^1 continuity. We would like to create patches that have C^2 or curvature continuity across their joins. In one dimension, we can create curvature continuous composite curves from a set of points using the method of cubic splines. Splines are functions that minimize the strain energy along the curve. They are historically called splines from the long thin strips that early builders used to approximate curves through a set of points. To create curvature continuous composite curves we will use Hermite interpolation between the sets of points, but we will specify positional, first derivative and second derivative continuity conditions at the adjacent knot points. For a cubic polynomial curve, we need 4 constraints to compute the 4 coefficients. Given a set of N points, we can define $N-1$ spans between each pair of adjacent points. If we fit a cubic polynomial to each span, we need a total of $4\cdot(N-1)$ constraints. Each of the $N-1$ curves has 2 positional constraints, for a total of $2\cdot(N-1)$ constraints. If we require continuity of first derivatives at the curve joins, that yields $N-2$ further constraints. Requiring second derivative continuity at the joins (which makes the curves curvature continuous) yields $N-2$ constraints also. There remain

$$4\cdot(N-1) - 2\cdot(N-1) - (N-2) - (N-2) = 2 \qquad \text{(A.22)}$$

two constraints before the set of composite curves is completely specified. Possible constraints that may be added are knowledge of the first or second derivatives at the first and last of the knots. If we can add these two extra constraints, then the composite curves are completely specified. Extending this idea to two dimensions, we start with a rectangular grid of knot points, $\mathbf{P}(m,n)$ $m=0,1,\ldots,M$ $n=0,1,\ldots,N$, that form $M\cdot N$ patches on the grid. We can create composite spline curves in each of the parametric directions such that the curves joining the knot points are curvature continuous. The extra conditions we need to specify are the tangents at each of the endpoints of the composite splines on the grid. Since we are also requiring the $M\cdot N$ patches to be curvature continuous across the joins, we need to interpolate the cross boundary tangent curves using the splining method. The two extra conditions imposed for this constraint are the cross derivatives (twists) at the corners of the knot grid. The information needed to create a series of curvature continuous patches is shown graphically in figure 6.2 and the algorithm that computes these patches is summarized in Faux and Pratt, pp. 224-225 [15].

Index